D0518746

BEGINNER'S GUIDE TO WEATHER FORECASTING

Streets of cumulus clouds on a summer's afternoon

BEGINNER'S GUIDE TO

Weather Forecasting

Stanley Wells F.R.Met.S.

PELHAM BOOKS

First published in Great Britain by Pelham Books Ltd
52 Bedford Square, London WC1B 3EF
1975

ISBN 0 7207 0819 2

Set and printed in Great Britain by
Tonbridge Printers Ltd, Peach Hall Works, Tonbridge, Kent
in Baskerville eleven on twelve point on paper supplied by
P. F. Bingham Ltd, and bound by James Burn
at Esher Surrey

To my wife, Joan

CONTENTS

ILLUSTRATIONS

LINE DRAWINGS

ACKNOWLEDGEMENTS

It is impossible to produce a book of this nature without being almost entirely indebted to everyone, professional or otherwise, who has in the past contributed to the skills and education of the subject, as well as to those who currently multiply the established practices with newly-acquired scientific and theoretical revelations which increase our understanding of meteorological phenomena.

In this connection I am pleased to acknowledge the wealth of study-material provided by exhibits in the Science Museum, London; the contribution of those whose works I have read over a period of many years; and, perhaps above all, the knowledge derived from over twenty years of lecturing in this subject to hundreds of young men who eventually took their places among the air crews and ground staffs of the Royal Air Force.

Furthermore, I am indebted to Messrs Negretti and Zambra Ltd for supplying the photographs of all the weather instruments reproduced herein, to the Controller of Her Majesty's Stationery Office for permission to reproduce selected parts of a Daily Weather Report of the British Meteorological Office, and for the photograph of a depression taken from a weather satellite.

The cloud photographs are reproduced by kind permission of Mr M. M. Rathore of Bracknell, and the Metmap form by the authority of the Royal Meteorological Society.

The remarkable photograph of our Earth taken from 22,300 miles in space is reproduced by permission of the National Oceanic and Atmospheric Administration, USA, and the description of meteorological satellites in Chapter VII also appears by permission of NOAA.

INTRODUCTION

Whatever else may pass one by unnoticed, no one can be unaware of the weather, for we are constantly reminded of its variable nature throughout the year by changes in its condition and, consequently, of our personal comfort.

According to these changes we are compelled to seek either warmer or cooler clothing, raincoats for showers and shelter from storms.

But apart from these personal indications of changes in the atmosphere, there are people to whom they are of even greater importance; for example, farmers may be prevented from planting by extremely cold weather, or fruit may not ripen owing to lack of sunshine. Rough weather at sea may prevent the fishermen reaching their fishing lanes, or from bringing back a satisfactory catch at the end of a stormy day.

Bad weather in the air may endanger aircraft or cause hazardous landing conditions, and so on into one inconvenience after another.

Few people have the opportunity of becoming deeply involved in weather-ways and weather forecasting, and a large majority are not aware that even a little understanding is within their grasp for the want of a little reading-time. It is, therefore, the object of this book to introduce the principles of meteorology to the non-expert, non-technical reader; the person interested in setting up his own garden weather station, or who merely feels a spark of curiosity about our much maligned weather. And from this offering the reader may learn from the signs provided by nature to estimate the probable weather reaction over the next several hours, or even days, by observing the clouds, the winds, the colours in the sky, air pressure, temperature and humidity; all simple but continuously changing facets of our weather systems, and presented here without the tedious mathematics and technical jargon of the professional meteorologist.

Obviously, if there happened to be a do-it-yourself method of *knowing* the weather, it is certain that people would be inconvenienced for no longer than it would take to learn the techniques of this new art. As it is, we must accept that we only know *about* the weather to the extent of the information available to us, and while advanced training and electronic aids can make us a little more knowledgeable than the traditional shepherd or seaman, at times, the weather expert is no nearer to predicting the changing conditions than is the City clerk at his office desk.

Nevertheless, meteorology is an interesting and intelligent hobby for the amateur, and since it does not necessarily insist that one leaves one's home in its pursuits, it must be ideally suited to the housewife with a little time on her hands and a little inclination to take up an unusual hobby. Nor is it necessary for the hobbyist to devote every day of the week to forecasting if it is not convenient to do so, for there is no reason why it should not be a week-end interest or rest-day relaxation for anyone who is unable to afford more regular attention.

Since ancient weather lore is inseparable from our connections with the ever changing weather, and since most people derive a certain amount of pleasure from these old sayings, there are relevant quotations throughout the book with which the reader may care to compare his own experiences of the stated conditions.

But just one small point should be mentioned – no matter how much you get to know about the weather, there will be times when you will be caught in the rain without an umbrella, or trapped in scorching sunshine wearing your raincoat.

'The weather is the eighth wonder of the World,
for the World is for ever wondering about it.'

Stanley Wells
London 1975

1/THE EARTH'S ATMOSPHERE AND BEYOND

We must consider the air as a fluid, for it carries certain characteristics of water when applied to bodies immersed in it, even to the extent of making them wet on occasions.

The air is a mixture of invisible gases which form a gaseous envelope, or bubble, around the Earth. This envelope is some 200–300 miles deep, submerging the Earth and isolating it from space.

The density of the envelope varies with increase of distance from the Earth's surface, being purest at its extreme height and becoming denser and heavier as it approaches sea level.

Towards the surface the atmosphere becomes more disturbed, turbulent and variable in character, giving a layer of air which is subject to storms, changes of wind, temperature, pressure and humidity.

The layer of air above this region is calmer and becomes more so as the air density decreases with height. The only storms are of an electrical nature, while high winds prevail at certain altitudes, and there are changes of temperature horizontally but not vertically.

Still further from the surface, air becomes less compressed and less dense, until there is no air at all and everything is a vacuum.

Contrary to the common belief of ancient times, air is not an element. It is composed mainly of oxygen and nitrogen with percentages of other gases, as shown below:

Nitrogen	78.04
Oxygen	20.99
Argon	0.937
Carbon dioxide	0.03
Hydrogen	0.0001
Neon	0.0012
Krypton	0.000005
Helium	0.0004
Xenon	0.0000006
Ozone	0.00014
Water vapour	varying amounts

Plants absorb some of the carbon dioxide and emit oxygen. Humans and animals absorb oxygen and emit carbon dioxide. Nature provides a system of maintaining these gases in an approximately even state of balance, except that the amount of water vapour is dependent on temperature variation.

In Industry the gases can be separated from each other and used for specialised purposes. Neon, for example, is used for illumination, and gives off a bright red glow. The addition of metallic mercury produces blue, while gold and white may be made by using helium and argon, and so on.

In the upper atmosphere there is a belt of air in which the gases are much the same as they are in the lower atmosphere except that it contains a small quantity of ozone. For this reason, and because of its importance, that belt of air is called the 'ozone layer'.

Although by comparison with the mixture of air that carries it, the ozone content is small, it serves to arrest the full force of the Sun's rays striking the Earth, thereby preserving our life as we know it.

Each layer of the atmosphere is known by a special name, giving three positive divisions which do not begin or end at sharply defined boundaries, but rather merge into one another in transitional zones (Fig 1).

The Troposphere

This is the lower layer, which extends to approximately 7 miles over the two poles and approximately 10 to 12 miles above the Equator. It is the area in which all our weather occurs, and is consequently the area with which forecasters are mainly concerned, although weather predictions are also made from studies of the upper air. It is also the area in which all general flying takes place.

At about 30,000 feet or higher, there is a narrow band of winds travelling at something like 100 knots; this is known as the Jet Stream.

Continuing upwards, there is a transitional zone leading to the layer above; this is called the *Tropopause* and varies in depth between a few hundred and a few thousand feet.

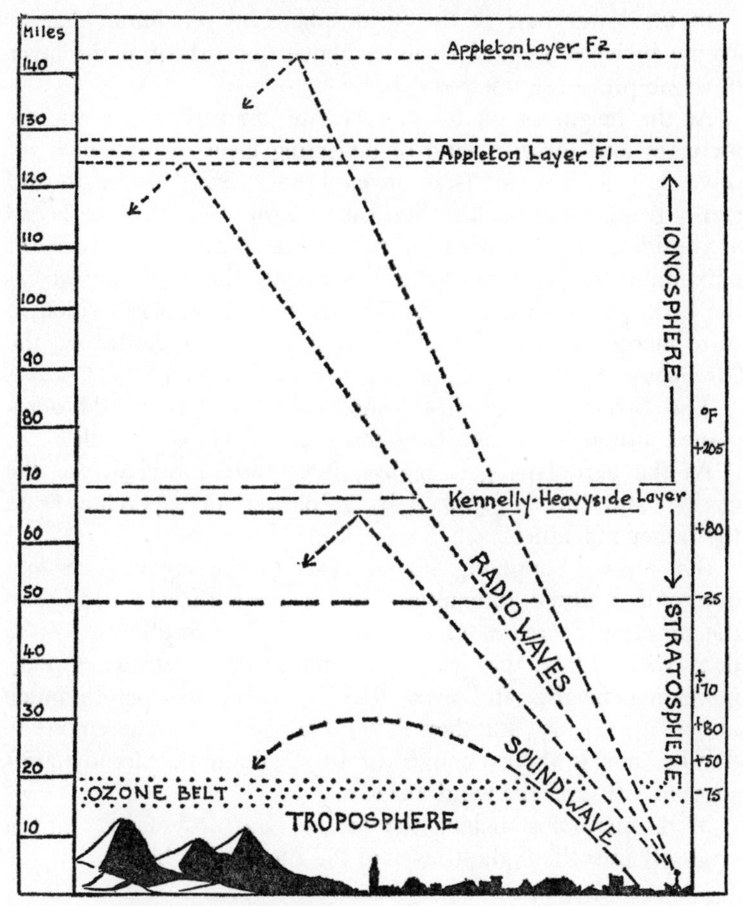

Fig. 1 *The principal layers of the atmosphere*

The Stratosphere

Here is a belt of almost cloudless, rarified air rising from the Tropopause and extending to about 50 miles above the surface.

There are few storms except for electrical ones, but there are high winds following directions which are chartable, rather like the charted sea currents and Trade winds of the surface air. The temperature is fairly constant with altitude but variable with horizontal distance.

In the lower part of the Stratosphere, its maximum density arising between 15 and 18 miles above the surface, is the layer of ozone previously mentioned.

At the height of 25 to 30 miles the temperature of the air increases and forms a kind of deflecting layer of air which receives any loud noises (such as explosions) from the Earth and gently bends them back so that they return to Earth to be heard at considerable distances from their point of origin. This area of maximum temperature corresponds with the layer known as the Stratopause, which trails off into the Mesophere, which in turn merges into the area of minimum temperature at the Mesopause.

The Stratosphere is just penetratable by pressurised aeroplanes, although balloons have reached a height of 13 miles.

As the aeroplane gets higher, it requires more air for the engine, and this is supplied by a 'blower' or supercharger in the larger machines.

But merely supplying forced air into the engine does not overcome altitude difficulties, for in these rarified conditions the airscrew decreases in efficiency as it has insufficient air to deal with. Only the jet engine increases in efficiency with increasing height, and even that is unable to operate much above 80,000 feet, for there is not enough air for the engine to handle, nor is there enough air to maintain the aerodynamic force of 'lift' over the wind surfaces.

With increasing height the air becomes less dense until it trails off into the highest layer of the Earth's bubble.

The Ionosphere

Occupying the depth of our atmosphere between about 50 to 300 miles up, the Ionosphere is almost airless and has secrets as yet barely explored by man.

Owing to the absence of air, aircraft with wings, airscrews or jets will not fly here, and it is therefore left to the rocket-propelled aircraft which derive their power from the reaction of the backward force of the fired rockets or jets. The rare atmosphere has been ionised by intense radiation from the Sun, so forming a conducting layer of electricity which bends medium and long-wave radio waves back to Earth so that they are received at distances which would otherwise be impossible, owing to the fact that they would simply shoot past the curvature of the Earth into space, and long-distance radio communication would thus never be possible. Only short-wave radio waves can penetrate the reflecting layers and carry on into outer space.

The layers responsible for this reflection were at one time clearly known after the names of the pioneers who discovered them – Kennelly-Heavyside layer at about 70 miles up, the Appleton layer F1 at about 130 miles up, and the Appleton layer F2 at about 140 miles up – but more recently they have been de-personalised and designated D, E and F.

Beyond the layers of the Earth's atmosphere lie the aerial wastes of space and outer space, the realm of weather satellites, space modules and an overpowering sun brilliantly shining from a black sky.

Change of sky colour with altitude

With increase of height from the Earth's surface, the various colours of the atmosphere become visible in their respective layers. From sea level, the sky appears light blue and changes approximately as follows:

5	miles	marine blue
6–7	miles	dark blue
8	miles	dark violet
11	miles	dark mauve
12–13	miles	dark violet-grey
13–14	miles	greyish black
15	miles	eternal night

Considering the blue sky that we are normally able to see in fair weather, we are at first surprised to learn that the Sun is really shining brilliantly in an almost perfectly black sky at all times.

Even the stars are always visible from about 13 miles up. Their appearance to us on Earth is made possible only when our position is turned away from the Sun into darkness. During daylight hours the Sun's rays are reflected and scattered by the molecules which compose the atmosphere, and as these molecules characteristically scatter blue light, the sky is completely spread with blueish illumination.

Lower, towards the surface, where dust particles increase their size and are larger than the higher molecules of the upper air, *all* the light is reflected and we experience the whiter effect at ground level, an effect more readily seen by looking towards the horizon, and one which we call haze.

The Sun

> 'The Sun is a-shining to welcome the morn,
> With a heigh ho come to the fair'

An enormous incandescent ball of fire, the Sun, from about 93,000,000 miles away, provides our Earth with heat and light, and is directly responsible for manufacturing our weather, although the explanation of this must occupy the remainder of this book, so intricate is the pattern of the Sun's action upon us.

If the Sun were a dark and dead world, our Earth would be bitterly cold and nothing would live upon it. With no heat whatever, all water would be in the nature of snow and ice, and the world a frozen waste.

Without heat it would not be possible for water vapour to ascend to the heavens and form cloud, therefore no rain would ever fall, and there would be no flowing rivers and no water to feed the dry, cold soil for the production of our food.

On the other hand, if the Sun were to shine continuously on the entire Earth, as it would if the Earth were still and flat, there would be no day or night and no seasons, and under the full intensity of the Sun the Earth would become a burnt crust.

By having land masses only, with no water or clouds, there would be no difference of atmospheric pressure, and consequently no weather, because this difference is brought about by land and sea masses having different heating properties and producing, by virtue of this, cold and hot air masses which, in moving from one area to another, produce the winds, clouds and rain.

Because the effects of the Sun are so important to us, man has developed a means of measuring its duration at various points on the Earth's surface, and all of us are familiar with the efforts of seaside holiday brochures to assure us that their town has registered a certain amount of sunshine taken on average.

The Sunshine Recorder

The first sunshine recorder was made by Campbell in 1853 and improved upon by Stokes in 1879. The photographic sunshine recorder was developed in 1884 by Jordan, whose father had demonstrated the principle in 1838.

There are two types of recorder, the burning type and the photographic type.

The first type, made in 1853, comprised a water flask mounted in the centre of a wooden bowl, the sunlight being focused on the interior of the bowl and burning a trace of its path as it moved across the sky. In 1887 a glass sphere replaced the flask, and the focused sunshine was caused to burn along the length of a graduated strip of paper.

The photographic recorder registers sunlight and ultra-violet light, both of which depend on the intensity of the Sun.

The glass sphere acts like a burning glass, focusing the sunlight on a strip of specially prepared card which is inserted into the curved metal plate behind the sphere. When the Sun shines, its course is traced by a burn line along the card which is graduated in hours. If there is only intermittent sunshine, the burn line is interrupted and the duration of the broken lines are then added together to obtain the total number of hours for which it received the Sun (Plate 2).

The Earth

'The big Earth is at the mercy of a little weather'

Because the Earth is round, or nearly so, the side which faces the Sun is said to be in daylight, and that concealed from it is in night; and owing to its curvature, the Sun's rays fall on different parts of the surface with different intensity depending upon the angle of the curvature.

At the Equator, where the Sun appears to be overhead, the Earth is hottest, and at the Poles, where the Sun appears to be at an angle, the Earth is coldest; it is by this difference of temperature and air pressure that the basis of our circulation of winds is to be found.

The Sun's rays reach full power in Central Canada and Central U.S.A. some 6 hours after they have brought Great Britain into daylight.

The Earth receives alternate periods of day and night by spinning from west to east and completing one revolution every 24 hours.

This means that objects at the Equator are carried through 24,902 miles in 24 hours at a speed of approximately 1000 miles per hour, the average speed of the rotating globe being 18½ miles per second, so you can see that we are already travelling at a very high speed before we even board that Inter-City train to Bristol or that jet-liner to Paris.

If Great Britain is at midnight, it is evident that it will be sunrise after six hours of revolution, and that after a further six hours it will be midday; six hours later comes the setting of the Sun, after which, in another six hours, it will be midnight again.

This arrangement allows nature and man to experience hours of rest and changes of temperature which would not otherwise come about, although modern-day lighting and heating engineers are trying very hard to eliminate these changes in nature which are our natural inheritance.

In additon to spinning, the Earth moves round the Sun once in twelve months, taking 365½ days to complete its orbit, and by so doing it creates seasons, assisted, of course, by the fact that the Earth is slightly tilted which brings the northern hemisphere near to the Sun in March and June, and furthest away in September and December. These are, respectively, spring, summer, autumn and winter.

In December the North Pole endures weeks of darkness. *March brings equal days and nights,* known as the Spring Equinox, June gives the North Pole weeks of daylight, and

September brings the equal days and nights in what is known as the Autumn Equinox.

21st March is the Vernal or Spring Equinox, and 22nd September the Autumn Equinox.

Spring officially begins on 21st March. Summer begins on 21st June. The Summer Solstice gives us the longest day. Autumn begins on 22nd September. Winter begins on 22nd December, and the Winter Solstice gives us the shortest day and longest night.

It is the rotation of the Earth which, in causing the rise and set of the Sun, is responsible for the difference in time factor of various parts of the Earth.

In 1884, the world was divided into 24 zones of time, fifteen degrees apart and one hour of time between each point; and time is accepted as travelling from right to left (east to west).

We are able to locate and refer to places on maps by means of lines of longitude and latitude. The northern and southern hemispheres are divided north and south by the Equator, an imaginary line midway between the Poles, and called a Great Circle.

Other circles are drawn at intervals parallel to it, the last encircling the Poles. These are parallels, or lines of latitude, and places are given as being on a line of latitude either north or south of the Equator, the Equator itself being 0 degrees.

Circles are also drawn perpendicularly to the Equator, passing through the Poles. These are called Meridians or lines of longitude, and give reference to place distance east and west of a given Meridian – that of Greenwich, England.

Measurements are given in degrees, minutes and seconds. A position is stated in respect of lines of longitude and latitude – thus the position of New York is 41°N. 74°W.

In terms of size, the Earth is 24,902 miles in circumference and 7926 miles in diameter at the Equator. The surface area is about 196,550,000 square miles, of which there is three times as much water as land. Exposed land masses account for 55,500,000 square miles, and water takes up the remaining 141,050,000 square miles.

Because the Earth is not a perfect sphere but is slightly flattened at the Poles, it bulges slightly at the Equator and therefore the diameter at the Equator is about 26 miles greater than the polar diameter.

In considering the tremendous speed at which the Earth is rotating, one may well enquire why it is that the globe is able to retain all the objects on its surface in a state of individual equilibrium, since the effects of centrifugal force tends to throw bodies outwards.

The force of gravity, which is responsible for attracting bodies to the Earth, is sufficient to counteract this centrifugal effect, thus keeping us closely attached to the surface.

Should the rotation speed be increased considerably, everything would suddenly be flung into space.

Erosion of the Earth's surface

Large-scale profile alterations in the land masses of the world are brought about in the course of centuries by weathering and erosion.

The land surfaces, hills and rocks and waterways, have since the beginning of time been gradually changing their shapes and courses as a result of deterioriation, crumbling and general breaking-up due to the destructive action of the weather.

Violent movements were originally responsible for throwing up some parts of the Earth above the level of others, so forming basins, plateaux and mountainous regions. It is likely that the surfaces of these upraised sections were mainly of a smooth, undulating nature, but centuries of extreme temperature changes, destruction by ice, earth movements from within and landslides from without, have caused the rough, rugged surfaces we see today.

High winds and rainfall wear away large areas of the upraised ground, and in winter the expansion of water turning to ice in the crevices and cracks split them apart and provide a passage for future rain water to begin its path of erosion.

So powerfully destructive is rain that it will penetrate the surface and begin erosion at depths of more than 200 feet. The natural caves formed in solid rock are often the result of millions of years of water percolating through these great depths.

The Moon

> 'Near full moon, a misty sunrise,
> Bodes fair weather and cloudless skies'

The Moon, reflecting the otherwise hidden rays of the Sun, gives us natural illumination on clear nights. It gives the illusion of being the same diameter as the Sun because, although it is so much smaller, it is a good deal nearer to us – only about 239,000 miles away as compared with the Sun's distance of 93,000,000 miles.

The diameter of the Moon is about 2,160 miles, against the Sun's 864,000 miles, and it accompanies the Earth as it travels round the Sun, revolving around the Earth in a regular orbit of its own.

The Moon's orbit around the Earth is elliptical, as a result of which the Moon is nearer the Earth at some times and further away at others. It takes about 27 days and 8 hours to travel round the Earth and exactly the same time to rotate once upon its own axis, so that we, at the centre of revolution, see only one hemisphere.

From the beginning of time until the Russian space probe Lunik 3 transmitted back pictures of the far side of the Moon in 1959, man had never seen the 'black' side of his nearest neighbour in space, although he has become familiar enough with its apparently changing shape as the month goes by.

The phases of the Moon are really only changes of appearance brought about by the fact that the Moon has no light of its own but reflects that of the Sun, and in moving round the Earth receives an amount of sunlight on different portions of its surface, thus appearing to us to change its shape.

The phases range from New Moon (not visible at all), through First Quarter (half illuminated) to Full Moon (wholly illuminated).

From Full Moon the sunlit portion of the face turned towards us slowly shrinks, through Last Quarter (half illuminated) to New Moon again.

The Moon is held in its orbit by the gravitational attraction of the Earth, its own gravitational pull being many times the weaker.

The effects of gravity on our tides

Twice during every 24 hours of the Earth's revolution, the tides come in and go out, rolling back and forth as the spinning globe is affected by the gravitational attraction of the Moon and the Sun.

Both the Moon and the Sun act like magnets, influencing the Earth to varying degrees, depending upon the relative positions of the three bodies in space.

By virtue of its more solid state, the land is not visibly affected by this changing magnetic force, but the water in its seas and oceans are caused to ebb and flow in what we know as tides.

The Spring Tides are brought about by the combined pull of the Sun and Moon when they are in one line with the Earth. This causes the high tide to be even higher than usual and, conversely, for the low tide to be lower than usual, and occurs twice during each month – when the Moon is new and when it is full.

Neap Tides result from the Sun and Moon exerting their forces in opposition to each other when the Moon is at its first quarter and at its last. At these times there is least difference between the level of the high and low tides.

The eclipses

Proof of the movement of the Earth and its Sun and Moon is provided by their eclipse.

The Moon, in revolving round the Earth, must at some time come between the Earth and the Sun, thus obscuring the sunlight and casting its shadow upon the Earth. This is called the eclipse of the Sun, or a solar eclipse (Fig 2).

Each month the Moon encircles the Earth, but it is seldom that the three bodies are in one direct line with each other; sometimes, in fact, only part of the Sun is covered by the Moon and therefore only part of the light is obscured, in which case it is called a partial eclipse.

The Sun, in illuminating the Earth, causes the Earth to cast a long conical shadow into space, in much the same way as a

Fig. 2 *Eclipse of the Sun*

Fig. 3 *Eclipse of the Moon*

person casts a shadow on the ground. The Moon, during its passage around the Earth, sometimes passes into this shadow and is obscured from view. This is called the eclipse of the Moon (Fig 3). If the Moon is not completely hidden it is only a partial eclipse.

The difference between the two is that in the solar eclipse, the shadow of the Moon is cast upon certain places on Earth and the effect is of a purely local nature, the amount of Sun visible depending upon the observer's position on the ground.

Those places which see the total eclipse of the Sun are on what is known as the line of totality, either side of that line being able to observe only a partial eclipse.

In the case of the lunar eclipse, the shadow of the Earth is cast upon the Moon and it is obscured from all observers at all parts of the Earth. The partial eclipse is when the Moon does not enter the entire shadow area and some of it remains visible.

By their knowledge of the regular movements of the three planets, through space and in relation to each other, astronomers can accurately predict the dates and times of all the eclipses.

In summing up the intention of this chapter, it is reasonable to say that we have established that the continuous changing of the Earth, Sun and Moon relative to each other gives us on Earth our continuously changing experience of night and day, weather, tides and season. We must now turn our attention to the air in which all our weather changes are taking place.

2/THE PRESSURE AND TEMPERATURE OF THE AIR

One might well ask why the gases which comprise our atmosphere do not rise from the envelope which they form and leave the Earth altogether.

It is plain to see that if we turn on the tap of a gas cooker, the escaping gas soon spreads across the room and does not remain close around the gas ring, as the atmosphere remains close around the Earth.

The reason is that, unlike the Earth, the gas ring has no gravitational force of its own. Gravitation draws together all the materials which make the Earth and retains them in the spherical shape with which we are all familiar, and in this fashion gravitation also influences the atoms of oxygen and nitrogen.

As these atoms are so much lighter than the Earth, and easily moved, they are drawn down towards it and are held there, being affected only by the movement of the Earth and the variations of pressure upon it.

While the Earth's gravitational field is exerting its powerful influence in order to retain the gases about its circumference, the gases themselves are exerting an equal opposite force in order to escape from its influence, so that at the Earth's surface, where the pull is strongest, the atoms are compressed very closely together, and therefore the air is more dense.

With increased distance from the surface, the force is exerting less influence on the gases and the atoms are able to move further apart, thus forming thinner, or less dense, air.

Consider that the air is made up of a series of layers, and that those layers nearer the surface are bearing the weight of those above them. The compression is high, and therefore the air is dense.

As height increases, the number of layers decreases, making the compression less for a given distance. The air becomes less dense owing to loss of weight.

This continues to happen until finally there is no air at all – just space.

If we had a pile of 8 books giving a total weight of 8 pounds, the lower book would have its pages compressed with the weight of those above, and would be supporting 7 pounds, assuming all the books to be the same weight. The third book in the pile would be supporting only 5 books and its leaves would be compressed by only 5 pounds of weight. The top book would not be supporting any others and so would not have its pages compressed at all – except by the weight of the atmosphere.

It is sometimes difficult to believe that the air around us has weight. We are not consciously called upon to exert any upward force before we can rise from a chair or pick up a book, but the reason we find it possible to lift a book from a table is that the air is pushing upwards away from the Earth with just as much force as gravity is exerting to pull the air down. The result is that an equal pressure exists all round all objects in the atmosphere.

As long as there is air all round the object we are able to move it, but try the following experiments for yourself and see what happens when there is no air on one side of the object to be moved.

Take a flat rubber disc, such as is sometimes sold for sink-stoppers, moisten the underside and drop it flat on an even surface. You will find that it takes a fair amount of strength to lift it, and that it will not budge until some small amount of air has crept underneath it.

The same thing can be done with a square piece of window leather soaked in water. Attach a piece of thin string to its centre, keeping the knot small, and seal the hole through which it passes with a glued-on patch of leather. Drop the wet leather squarely upon a smooth surface, and then try to lift it off again. One is more likely to tear the string through the leather than to part the leather from the surface.

The explanation is that, in dropping the wet leather on to the smooth surface, you have exhausted the air from beneath it and there is no longer any upward pressure. All the force is now exerted downwards by the atmosphere above the leather.

A suction cup is effectively based on the same principle of excluding air from between it and the surface to which it adheres.

If our bodies were not constructed so that the gases and fluids within pressed outwards with equal force to balance the pressure outside, we should be crushed by the weight of the air of almost 15 tons, which is the approximate pressure exerted on the average adult.

The total weight of the atmosphere on the Earth is nearly 6000 million million tons. The column of air rising directly above each one of us weighs about 1 ton. The downwards force of the atmosphere is about 14.7 pounds per square inch at sea level, depending a little on latitude and temperature. This decreases to about 7 pounds at 20,000 feet, for the higher we go in the atmosphere the lower the pressure, as there is less air above us; and 30 miles up, the pressure is about 1/1000 of that at sea level.

To prove further that the atmosphere really does press upwards with such force, we shall attempt another experiment. You will require a drinking glass and a piece of stiff card large enough to cover the mouth of the glass and overlap it by about two inches all round.

Take these to the sink (just in case of accidents) and fill the glass with water. Put the piece of card firmly and squarely down over the mouth of the glass and, holding the bottom of the glass in one hand while pressing the card against the mouth with the other, turn the whole lot upside down. Now remove the hand holding the card and observe. The water does not spill into the sink. The card remains in place by virtue of the fact that the weight of the water is less than the upward pressure of the air, and therefore the air pressure, in exerting an upwards force and keeping the card in place, is at the same time supporting the weight of the water. This will continue to be the state of equilibrium until the card becomes saturated with water and finally allows air to come between itself and the water, whereupon it will fall away.

We have just demonstrated how air is exerting an upward force. We will now continue in this experimental mood in order to show how the air presses downwards as well.

All that is required is a U-tube from a chemistry set and a small amount of water.

Pour the water into the tube to about the half-way level and you will see that the surface of the water in both sides of the

U is at the same level. This is because the atmosphere is exerting an equal force on both surfaces.

If we had a pump which would seal off one side of the tube and exhaust the air from it, we would see the water in that side of the tube rising until it reached the sealed top, and the water level in the unsealed side would sink. The reason for this is that there is no atmospheric pressure in the exhausted side of the tube, and therefore all the weight of the atmosphere is pressing down on the open side. This pressure is greater than the weight of the water, so that the column of water is forced up into the evacuated side of the U.

However, the absence of a mechanical pump need not deter us; there is a simple way out of this difficulty. All you have to do is to put your mouth over the end of the tube and suck the air from it until the water level rises to touch your lips. You have become a human vacuum pump and the weight of the atmosphere has pressed the column of water up into the airless section of the tube.

As a final experiment in expounding this particular theory, let us again pour water into the tube so that it reaches the half-way mark in both sides, when we know the pressure to be equal. This time, fit a cork into one end of the tube and tilt it towards the corked-up side so that the water touches the cork, and whatever air may have been in that side flows down into the open tube. Observe that the water in the corked side will remain supported in the column by the weight of the atmosphere on the open side. This proves that the atmosphere may be used to balance a measure of liquid in a tube.

We should therefore be able to construct an instrument along these lines which could be calibrated to read the changes in the pressure of different types of air as they pass over the apparatus.

If the air is cold and heavy we would expect the column of water to rise, owing to the greater pressure on the one side of the tube, and we could mark the sealed side of the tube in inches and say that the column of water rose to a height of so many inches when that area of cold air passed overhead.

If the air changed to a mass which was warm and light, we would expect the pressure on the surface of the open tube to be less, and this would allow the column of water in the sealed tube to fall an inch or so. By this means we could measure the varying pressure all the year round.

Water glasses, working on this principle, were in use for many years before the barometer was invented, and an engraving in the British Museum, dated 1631, illustrates the making and use of a weather glass. The instructions on it tell us that 1. the water ascendeth with cold and decendeth with heat, 2. if in 6 to 8 hours the water drops a degree or more it will sure to rain within 12 hours after, and 3. by diligent observation one may foretell frost, snow and foul weather.

However, water is an inconvenient medium as it is subject to rapid evaporation and will easily freeze, so that the standard barometer uses a column of mercury with which to balance the atmosphere.

The Mercury Barometer

'When the glass falls low,
Prepare for a blow;
When it rises high,
Let all your kites fly'

Because of the irregular heating of the Earth, the atmospheric pressure at any one place does not remain constant. The existence of atmospheric pressure was demonstrated by Torricelli in 1644 when he developed the mercury barometer; and the modern barometer is essentially of the same form as the original.

The standard Kew Pattern Station barometer was designed by the Director of Kew Observatory, John Welsh, in 1854; those used today in Meteorological Offices are of this design but with minor modifications (plate 1).

The barometer consists essentially of a glass tube 33 inches long, sealed at one end and almost filled with mercury. This tube is inverted into a cistern of mercury so that the open end is below the level of the mercury in the cistern, and the sealed end is uppermost.

The column of mercury will assume a position where it just balances the weight of the atmosphere, and will stand at a height of approximately 30 inches. This height varies with changes of air pressure; when there is a decrease in air pressure the mercury falls in the tube, and when there is an increase, it rises, but it will maintain about 30 inches when the pressure is of average normality.

So, then, if the tube has a bore of 1 inch square, and the mercury stands at 30 inches, the weight of the mercury column is 14.7 pounds, proving to us that the air is pressing down with a similar force on every square inch of the Earth's surface. A difference of 1 inch in the reading is equal to almost $\frac{1}{2}$ pound weight of air.

It is these changes in weight and their effect upon the barometer which enables us to make certain observations concerning the expected weather.

As it is not convenient to construct a barometer with an exposed cistern, the entire apparatus is enclosed tidily in a metal casing and provided with a calibrated scale from which the pressure is read.

Although most domestic barometers are scaled in inches of mercury, from just below 28, through 29, 30 to just above 31, this is an inconvenient method to use since changes in pressure are often quite small (in the order of sixteenths of an inch can be expected), and so meteorologists prefer to use the unit of the bar as a measure of pressure rather than inches as a measure of mercurial height.

1 bar equals 1000 millibars or 29.531 inches of mercury. The average normal pressure of 14.7 pounds, represents 1013.2 millibars (mbs), and this represents a column of air 1 inch square and reaching right to the top of the atmosphere.

In the British Isles, as in all temperate latitudes, pressure changes are irregular and erratic due to frequent advancing pressure systems, while in tropical latitudes the change is small except for diurnal variations.

The maximum pressure occurs at about 1000 hours and 2000 hours (10 a.m. and 8 p.m.), and the minimum at 0400 hours and 1600 hours (4 a.m. and 4 p.m.) local time.

The height of the mercury is dependent upon:

a. atmospheric pressure
b. temperature
c. latitude
d. altitude

To clarify these points it is necessary to point out that air in contact with a warm surface will itself become warmer than the surrounding air and in consequence become less dense and more buoyant, rising to higher regions and being replaced by cooler air as it does so. An area of warmed air, being less dense than cold air, will be indicated by a rise in the barometric pressure reading. In latitude 45°, at a temperature of 32°F at sea level, the pressure will read 29.531 inches, or 1000 millibars. Up to 6000 feet above sea level, a difference of 1 millibar represents a difference of 30 feet, while a change of 1 millibar represents a difference of 50 feet through higher levels.

Because there is a considerable difference in pressure readings according to the height of the barometer above sea level, it is necessary to make corrections to the setting of the instrument in order to compensate for its height. For example, if the barometer is taken from sea level to the top of a mountain, it has travelled the height of the mountain above its place of origin and will therefore measure the weight of the column of air immediately above. An addition of something in the region of 1 millibar for every 30 feet above sea level will have to be made to obtain a corrected reading.

The Aneroid Barometer

The most familiar barometer in domestic use is the aneroid type, which, since it was found to be most suited to sea-going vessels, is also known as the Marine type (Plate 3).

The mechanism consists of a small corrugated drum which is exhausted of air, and is sensitive to changes of pressure (expansion and contraction) on its outer surfaces. The variations of its movements are amplified by a system of levers, operating a pointer which moves around a calibrated dial. The dial is marked in inches or millimeters of mercury, or millibars, or both.

The aneroid should be frequently checked against the standard pattern barometer and adjustments made according to any inaccuracy in the reading. Adjustment for the pointer is made by turning a deep-set screw at the rear of the container.

Positioning the barometer

The barometer should be hung in a cool position of even temperature, and sheltered from draught and wet. Readings from its dial or scale should be noted at regular intervals and observations made from personal experience and knowledge – not by such indications as 'stormy', 'wet' and 'fine'. The barometer alone cannot forecast conditions; we must apply knowledge of the wind, cloud, temperature and humidity before we can hint at the probable weather conditions.

The Barograph

The barograph is an instrument which produces a continuous record of atmospheric pressures during a given period of one week.

It consists, broadly speaking, of an aneroid barometer mechanism connected to an inked nib in place of a dial indicator, and the nib is caused to trace a line over a slowly moving chart fitted to a drum which is rotated by clockwork (Plate 4).

The pressure readings are taken off against a scale printed on the chart, and the pressure changes over the whole of the week may be consulted in this way. The chart thus provided is called a barogram.

Barometer indications in forecastings

'Fast runs the ant as the mercury rises'

The difficulty of setting down hard and fast rules must be appreciated, since the individual forecaster is concerned only with his local conditions, while the professional is concerned with the entire hemisphere, and the barometer alone cannot provide all the answers. The instrument is sensitive to changes of wind and consequently of pressure, while the thermometer is concerned with whether the origin of the wind lies in the cold or hot regions of the hemisphere, and the hygrometer is concerned with the moisture content of the air. The individual must concern himself with all of these factors as well as with the appearance of the sky, the condition of the clouds and wind direction before he can attempt a forecast.

However, the following indications based on barometer readings should assist during practical experiences, and it will be found possible to make several additions of one's own as familiarity with local weather conditions become apparent during study.

Pressure variations between 980 and 1030 millibars are fairly common for the British Isles and provide our average readings, while a low reading would be in the region of 950 millibars, and a high reading would be in the region of 1050 millibars.

Indicator movements are, more often than not, quite small and likely to be missed by the careless observer, but it is the change rather than the weather indicated that is important, which is why the professional meteorologist does not have 'dry', 'wet' and 'fine' inscribed on his Kew pattern mercury barometer.

It can be raining heavily when the arrow of the aneroid in the hall points to 'dry', or it can be hot and sunny when it points to 'wet'. One cannot always assume that a fine day will ensue if the reading is high, or that a wet day will follow if it is low.

What is important is its behaviour over the past few hours, and the smallest change is of significance in detecting the possible weather responsible for it.

One Sunday morning you may notice a reading of 1019 millibars (30.1 inches) at about 7 o'clock, and note that the arrow is in the 'Fair' section of the dial. This, taken at face value may be misleading because, owing to the inability of some aneroids to indicate immediate changes through the lever mechanism, the reading may not be accurate. A gentle tap on the casework may then show a movement of the pointer towards 1016 millibars (30 inches), indicating a falling reading and the approach of deterioriating weather. By the time you have decided to clean the car for the week, the rain is falling heavily and the barometer has entered the 'Change' section.

On another occasion you may observe a reading of about 999 millibars (29.4 inches) with the arrow indicating in the

Change/Rain section. A gentle tap on the casework may cause the arrow to jump another point or so to 1002 millibars (29.6 inches). Subsequent readings show a slow rise, and before long there is sunshine, or at least a fairly good afternoon, after a dull, cloudy morning.

The first example indicates the approach of a depression, which will probably take all day to pass over. The second example indicates that a night depression has passed over and is followed by an area of brighter weather.

But neither condition was obvious from the face value reading of the barometer.

Barometers are fitted with an adjustable indicator which can be set, independently of the main indicator to the reading at the time of observation. This indicator remains in position until re-adjusted at the next observation, thereby providing a point from which to see whether the main indicator is rising or falling.

Broadly speaking, a high reading will stand for fair to good weather and a low reading will be given for poor weather, subject to the variations shown by the indicators. For example, if the barometer remains steady at any particular position, it is reasonable to expect the conditions to prevail.

Slow movement, either higher or lower, will tell us that conditions are likely to continue settled. Sharp movements herald unsettled weather.

When the rise is from low to high, conditions are expected to be more favourable than a sharp fall from high to low.

Frequent 'jerky' movements about a given position are characteristic of unsettled weather, and it will probably be rainy or windy.

Any change in conditions that takes place when the wind is in the east is apparent much sooner than when the changes occur with the wind in the west.

The behaviour of the wind is important in these changes; for instance, we can experience a fine day with a low but rising barometer if the wind has already veered from south-west to north-west, and a wet day with a high but falling barometer if the wind has already backed from north-west to south-west.

The readings will be low in warm air and high in cold air.

A fall in the barometer may herald a spell of dull, rainy weather, while a rise may indicate dry, fine weather.

The name of Admiral Fitzroy (1805–65), Director of the Meteorological Office from 1855–65, is largely associated with the finely-made siphon barometer, which also included on its mounting board a thermometer and storm glass, together with extensive remarks concerning the reading of the weather from the rising or falling of the barometer.

The following guide is based on his and other observations, and it is of interest to note how wind and temperature are closely related to behaviour of the barometer, and how necessary it is to consider factors other than barometric readings when preparing a forecast.

Barometer indications when rising

a. A long, steady rise in summer indicates fair, settled weather and probably a hot spell.
b. A steady rise with dry air and falling temperature indicates wind from the northerly quarter.
c. A long, steady rise in winter may mean fine, cold, frosty weather and possibly fog. This depends upon the location of the centre and wind direction. If there has been wind and rain, such a rise may mean less wind and less rain.
d. In wet weather, a high reading maintained over a period may indicate the coming of fine weather. If the rise is sudden from low to high, the coming fine spell may not last long.
e. A rise with moist air and low temperature may bring wind and rain from the north.
f. A rapid rise – unsettled. The barometer rises higher for north and east winds than it does for south and west winds.
g. A rising barometer and rising thermometer with a dry southerly wind may bring a fine spell.
h. A very slow rise from low to high usually brings dry weather with high winds.
i. If the reading, having been at average for some time, is rising against a falling thermometer and with dry air, the indications may be for northerly winds or less winds. If there has been precipitation it may indicate less rain or less snow.
j. A barometer that is rising after having been below average

for some time may indicate less wind or a change to north, or less rain.

k. A rise following a period of very low readings in the region of 980 millibars may indicate heavy squalls from the northerly quarter, or strong winds.

l. A slow rise should bring settled weather.

m. Continued steadiness in dry air indicates very fine weather.

n. 'If the barometer and thermometer both rise together, it is a very sure sign of coming fine weather.'

Barometer indications when falling

a. A fall with rising temperature and increased humidity indicates wind and rain from the south-east, south-west or south.

b. A fall with low temperature indicates snow and rain.

c. A sudden fall at low temperature with a west wind may mean heavy gales from the north-west, north or north-east.

d. A steady fall indicates unsettled or wet weather in all seasons.

e. A slow fall from high to low indicates wet weather with little winds.

f. A rapid fall indicates sudden rain or snow, or high winds, or both.

g. A fall with rising temperature and wind backing from the north-west heralds an approaching gale.

h. A fall in hot weather indicates thunder. A sudden fall indicates high winds.

i. Falling during high winds from the south through to the west means an increasing storm.

j. A sudden fall during a westerly wind indicates a storm from the north-west, north or north-east.

k. When the barometer falls against a low thermometer it may snow.

l. In summer, a storm can be expected if the barometer falls sharply. If it does not rise again after the storm, there will be several days of unsettled weather.

m. 'Long foretold, long last; short notice, soon past.'

It is of considerable interest to the local forecaster to make his own observations over a long period and make additions to these indications.

Since we have seen that temperature is closely linked with air density, it is now time to proceed to the next step in our studies.

Temperature zones of the Earth

As we have seen, temperature, the sensation of heat and cold, is a continuously changing factor produced by the daily rotation of the Earth and its annual path around the Sun, so ensuring that no areas of the Earth are either burned or frozen beyond certain limits where they would become useless.

The Earth's surface may be divided into five temperature zones. At the Equator, or the 'torrid zone', the heat radiating from the ground raises the temperature of the air and produces a belt or bubble of warm air which moves slowly upwards and is, by virtue of its warmth, less dense than the air at the Poles.

Between latitudes 25° and 70° north and south of the Equator lie the 'temperate zones', and at each Pole are the 'frigid zones'.

Basically, our circulation arises from this difference in temperature, and consequently of pressure, in these zones.

The rising warm air at the Equator makes its way north and south through the temperate zones towards the Poles, slowly cooling and falling as it travels.

At the Poles, where the air is naturally cold and more dense than that of the Equator, there is a slow flow of air towards the torrid zone moving underneath the warm air masses in order to equalise the distribution of air as it leaves the Equator.

Between the two hemispheres, where the rising air departs one way bound for the North Pole and the opposite way bound for the South Pole, there is a belt of relatively calm air known as the Doldrums, an unpopular position with the old-time sailors whose ships were often becalmed there for weeks on end.

So it is that as these temperature areas are caused to move across the Earth, we detect the changing types of air by means of the thermometer.

Temperature

The temperature of the air varies from day to day and even from hour to hour. As well as this, there are long-term variations brought about by the changing seasons, and the cooling and warming of the Earth caused by the cycle of day and night.

Contrary to common belief, the heat we experience in the summer months is not produced by the rays of the Sun warming the surrounding air, but by the process of radiation from the Earth's surface.

The amount of heating by direct exposure of the air to the Sun is so small that it may be excluded from our consideration.

What happens is this. Sun rays, which are high-frequency radiations, penetrate the atmosphere and strike the Earth. In summer months the response to heating is rapid and in a short time the Earth becomes warm and then hot, but as it is not possible for the Earth to absorb all the heat it receives, it begins to radiate a large percentage of it back into the air in the form of long-wave radiation, as heat waves are much longer than light rays, and will heat the air at the surface. Since the density of the air is greatest near the surface, much of the radiation will be absorbed by it, causing the temperature to rise.

The warmed air will, by virtue of the heat, become lighter, or less dense, and will rise, allowing another mass of cooler air to take its place and in turn to become heated and to rise, so that there is a continual changing of the air, although in hot conditions we are not always aware of this movement.

When the sky is clear in spring and winter, the Earth is able to radiate all its heat into the atmosphere and so the temperature may be found at a fairly low mark by evening. This is assisted by the fact that the heating effect is comparatively small owing to the season and the low position of the Sun.

When there is a good deal of low cloud about, the heat is unable to escape from the surface layers of the atmosphere and the temperature may be expected to increase or remain steady.

The coldest months in England are January and February, although when the south wind blows, the thermometer may rise to 60°F or so.

The hottest months are July and August, when an average of 90°F may be expected.

The coldest place on Earth is Verkhoyansk in Siberia, where the temperature falls to 90° below zero, giving 122° of frost.

The hottest place on Earth is Death Valley, U.S.A., with a temperature of 132° in the shade.

The surface of the Sun is said to be 70,000,000°F.

The temperature of the human body in good health is approximately 98.4°F.

The interchange of temperature between bodies takes place by conduction, convection and radiation, and it is important to understand the difference.

Conduction of heat

Heat is conducted from particle to particle, and in terms of meteorology we may consider the propagation of the changes of temperature downwards through the earth from the surface as it is heated during the day.

Dense soils are better conductors of heat than loose soils, so that light soils are subject to high temperatures than are heavy ones.

Heat flows, or is conducted, from the body having the greater heat to the body having the lesser heat, in an attempt to even the balance, and since some bodies absorb heat better than others, there are different degrees of heating by different bodies.

Damp air is a better conductor than dry air, and consequently damp air feels colder than dry air of the same temperature. A marble table feels colder than a wooden chair, though both are at the same temperature.

At the breaking of a severe frost, when the temperature has risen and the air becomes moister, the weather can feel more chilly than when the temperature was below zero.

Snow is composed of crystals mixed with a large amount of air and is therefore one of the worst conductors of heat, but it serves to prevent the escape of heat from the earth to the air and limits the depth to which severe frost may penetrate, so protecting the soil and plants from damage.

Water is a good conductor of heat and becomes heated by the Sun's rays to a considerable depth during long exposure.

Convection of heat

Fluids and gases are bad conductors of heat, which is why the air is very slow to receive direct heat from the Sun, but they may be quickly heated by a process of convection by the circulation of their particles. Take, for example, a saucepan of water on a gas ring. When the heat is applied to the bottom of the saucepan, the metal is first heated and then transfers some of that heat to the lower depth of water in contact with its interior. The particles at the bottom are heated and become lighter, ascending to the surface, while other particles of cooler water descend to the bottom to receive their share of the heat. In this way two currents are formed, the hotter ascending through the centre and the colder descending down the sides of the saucepan, a process which continues until all the water has reached the same temperature, whereupon it tries to escape from the surface and causes the bubbles we see when the liquid is boiling.

The daily fluctuations of temperature on the surface of the Earth cause convection on an extensive scale all over the world, and the result is found in ascending and descending currents of air, warm being replaced by cool, cool being heated and rising, which is what happens when the Sun's rays heat the ground.

At the tropical regions of the Earth, the air is rapidly and excessively heated, and it immediately ascends and flows off towards the Poles, making way for the cold air from the Poles to flow in and take its place, so giving us the polar and equatorial currents which form part of the entire circulation of our atmosphere.

Radiation of heat

If we sit in the summer sunshine for very long, our skin begins to 'cook' and we can be blistered by the heat of radiation. When we stand in front of an open fire we receive some of its heat and may have to move away if it becomes too hot.

This kind of heating is called radiation and is an interchange of heat which is constantly going on among bodies freely exposed to each other.

When an area of the Earth's surface is turned towards the Sun, it will receive more heat than it can radiate, but when it is turned away from the Sun that area can radiate more heat than it receives.

This is a cycle which is continually in process, night to day, season to season, and it provides us with a division which we can call solar radiation and terrestrial radiation, the former being the amount and effect of solar radiation which reaches the surface of the earth, and the latter being the amount and effect of the Earth re-radiating its absorbed heat back into space.

The effect of temperature on sound

'A good hearing day is a sign of wet'

It is of interest to note, without lengthy discussion, that temperature effects the velocity of sound. In air at 60°F, sound travels at 1120 feet per second, or 12½ miles per minute.

In salt water, the speed at 36°F is 4,840 feet per second, but increase the temperature up to 45°F and the velocity of sound becomes 4,850 feet per second.

In ordinary water at 27°F, the velocity of sound is 4794 feet per second, but lower the temperature to 15°F and it becomes 4780 feet per second.

Factors affecting temperature

Temperature is not a simple occurrence relying on a single cause for its effect, but is influenced by several factors about which we may be quite brief.

Altitude

Air density decreases steadily up to about 7 miles and at that height receives no radiation from the earth, so that the temperature decreases with increase of height progressively from the surface throughout the Troposphere. This decrease is at the rate of about 3°F per 1000 feet, but at the Tropopause, the boundary between the Troposphere and the Stratosphere, the temperature stops falling and produces a band of warm air.

Latitude

Broadly speaking, temperature decreases with increase of latitude at the rate of 1°F per 1° of latitude.

Owing to the difference of the incidence of the Sun's rays, temperature is usually high in the tropics and low in the polar regions, and is considerably modified by the results of its passage over land or water areas, where it either becomes dryer or moister respectively.

Season

Since the seasons affect the duration of daylight and the incidence of the Sun's rays, they affect the amount of heat received also.

Land and sea distribution

Land is subjected to a larger range of temperatures than is the sea, but while the sea will act as a reflector of the rays, the land absorbs the heat and radiates it more readily than water.

Water requires four times as much heat to raise its temperature as does soil.

Prevailing wind

Since the wind, or air movement, is responsible for bringing air from cold or warm areas, it will modify local conditions considerably and is therefore an important factor in its effect upon temperature.

Cloud amount

Where there is a large cloud amount, it will prevent the Earth radiating its heat back into space, and will therefore prevent cooling by radiation. As a layer of cloud will reflect a considerable proportion of the Sun's heat back into space, it will consequently control the amount of radiation received, so that a cloudy day following a cloudless, sunny day may mean a reduction in temperature.

Water vapour

Ever present in the air in small or large quantities, water vapour, because of evaporation, is most prevalent over water areas and will exercise a blanketing effect which reduces the rate of heating.

It will absorb heat and reduce free radiation at night, which is often responsible for some of those warm, moist nights.

Land surface

The type of surface upon which the Sun shines will largely control the amount of heating. Hard ground heats faster than soft ground; fields of short grass heat faster than fields of long grass; ploughed fields heat faster than grassland.

Snow reflects heat back into the atmosphere and retains the warmth of the earth which it covers.

Obstructions such as hill and mountain ranges, trees and buildings, will cause a shaded area which will cool the adjacent areas.

From this information it emerges that temperature changes are the result of long-range as well as short-range and local conditions. Therefore, the evidence given by the thermometer readings should be regarded with the same care as those of the barometer.

Thermometers

> 'A sudden increase in the temperature of the air sometimes denotes rain; and again a sudden change to cold sometimes forebodes the same thing.'
>
> *Bacon*

Galileo, in 1592, was the first to measure temperature with his thermometer. This instrument was subsequently improved upon by the Accademia del Cimento, Fahrenheit, Reaumur, Celsius and De Luc.

In the late eighteenth century, Cavendish, Six and Rutherford made thermometers which registered the highest and lowest temperatures over a given period.

It is the function of the thermometer to measure the degree of hotness or coldness of a mass or body. The medical thermometer, inserted beneath the tongue, will measure the temperature of the body.

The meteorological application of the thermometer is for measuring the temperature of the atmosphere. These readings are usually expressed as being 'in the shade', unless it is specifically required to obtain readings 'in the Sun', in which case a specially prepared 'black bulb' thermometer is used.

The most common type of thermometer consists of a glass tube of fine bore with a cylindrical or spherical bulb at the lower end, the top being sealed. The bulb contains mercury which, when heated, will expand, thereby filling the bulb until it rises through the bore of the tube to a height determined by the amount of expansion (Plate 5).

When exposed to lower temperatures, the mercury contracts and tends to fit itself back into the bulb, thus registering lower readings depending upon the amount of cold, and therefore contraction, to which it is exposed.

As mercury freezes at about −37°F, alcohol is used for those instruments intended for registering extra low temperatures.

Two scales of reference are in general use today, the Fahrenheit and the Centigrade, and although there is a marked effort to convert us all to the latter, the temperatures in this book are, for the sake of simplicity, given in Fahrenheit, since this is still the most popularly understood scale in domestic use.

The Fahrenheit scale

The zero point on this scale is the temperature of a mixture of ice and salt (this being colder than pure ice). Next, the inventor, Fahrenheit, chose the temperature of his own body and marked the mercury level at 96°. The entire scale is based on the freezing point of water at 32°, separated by 180° from the boiling point of water at 212°.

The Centigrade scale

Anders Celsius provided his Centigrade scale with 100°. He immersed the bulb of the mercury into a mixture of ice and water and marked the mercury level reached as 0°. The bulb was then held in a jet of steam from boiling water, and the height to which the mercury rose was called 100°.

With these two scales in use it is necessary to indicate F after a Fahrenheit reading and C after a Centigrade reading i.e. 122°F and 50°C.

Comparison and conversion of the scales

1°F=5/9°C, and conversely
1°C=9/5°F.

To convert Centigrade to Fahrenheit, multiply the reading by 9, divide the result by 5 and add 32.

i.e. 50°C×9=450÷5=90+32=122°F.

To convert Fahrenheit to Centigrade, subtract 32 from the reading, multiply by 5 and divide by 9.

i.e. 122°F−32=90×5=450÷9=50°C.

By comparing the two working methods, it will be seen that one is just the reverse procedure of the other, but taken that 1 degree Centigrade is equal to 1.8 degrees Fahrenheit, and that 1 degree Fahrenheit is equal to 0.56 degrees Centigrade, the following table provides a ready reference comparison for a wide selection of temperatures.

Conversion Table

°C	°F	°C	°F
40	104	5	41
39	102	4	39
38	100	3	37
37	99	2	36
36	97	1	34
35	95		
34	93	0	32
33	91	−1	30
32	90	−2	28

Conversion Table

C°	F°	C°	F°
31	88	-3	27
		-4	25
30	86	-5	23
29	84	-6	21
28	82	-7	19
27	81	-8	18
26	79	-9	16
25	77		
24	75	-10	14
23	73	-11	12
22	72	-12	10
21	70	-13	9
		-14	7
20	68	-15	5
19	66	-16	3
18	64	-17	1
17	63	-18	0
16	61	-19	-2
15	59		
14	57	-20	-4
13	55	-21	-6
12	54	-22	-8
11	52	-23	-9
		-24	-11
10	50	-25	-13
9	48	-26	-15
8	46	-27	-17
7	45	-28	-18
6	43	-29	-20

Maximum and minimum thermometers

The device used for registering the highest and lowest temperatures over a given period is called the 'maximum and minimum thermometer', a self-registering U-shaped thermometer named after James Six who invented this design in 1872 (Plate 6).

As it is not possible to maintain a continuous round-the-clock observation on a single thermometer, it is obvious that a self-registering instrument is invaluable in giving us the maximum and minimum temperatures experienced in 24 hours. Officially, the most suitable hours for observations to be taken are 6 a.m., midday, 6 p.m. and midnight.

In winter, the maximum temperature occurs between 1 and 2 p.m. and the minimum between 6 and 7 a.m. In summer, the maximum occurs between 4 and 5 p.m. and the minimum between 3 and 4 a.m.

The instrument is made as a U-formed mercury thermometer, showing two scales exactly parallel with each other, and with a bulb at each end.

The left-hand side shows minimum readings and the right-hand side shows maximum readings. These are determined by means of a short steel wire, looking rather like a blunt tacking pin, situated at each end of the column of mercury.

With a decrease of temperature, the mercury is caused to push the index up into the left-hand arm of the U-tube. With an increase of temperature, the mercury is caused to retreat and expand into the right-hand arm of the tube.

As the mercury drops away the index is left in place, enabling one to read the minimum temperature when convenient. In the meantime, as the mercury rises in the right-hand arm, the index is carried upwards to register the maximum temperature to be reached during the observation period.

The indices are reset by drawing a magnet downwards on the outside of the tube until they reach the surface of the mercury once again.

For professional purposes the Six's Thermometer is not accurate enough and does not fulfil all the requirements of the full-time weather man, who has four individual thermometers mounted in a special housing, known as a Stevenson or Bilham screen, standing 4 feet above the ground, with louvered sides to allow for free circulation of the air.

One thermometer has its bulb encased in wet muslin, and is used in conjunction with its neighbouring air temperature thermometer to provide a reading known as the 'dew point', of which we shall speak later on. The other two occupants of the screen are the minimum and the maximum thermometers, both mounted almost horizontally, the former having an alcohol column and the latter a mercury column.

For convenience in domestic use, it is usual to combine two or more weather instruments in the banjo-shaped piece of furniture that is usually hung in the hall, sometimes in the draught from the front door, sometimes immediately above the central heating radiator, and sometimes on a damp outside wall. When consulting an indoor collection of instruments it is worth noting that they will register the conditions indoors, and that a similar set of instruments out of doors, where the weather occurs, will provide different readings, particularly in respect of temperature and humidity.

Sounding the upper air by radio sonde

The weather map gives us a good picture of conditions near the surface of the Earth, and these are the conditions with which our weather instruments are concerned. However, our atmosphere extends for many miles and is divided for purposes of identification into the layers previously described.

The movement and condition of the air in these regions cannot be accurately assessed from the ground, and it is not convenient to provide for frequent observations to be made by aeroplane.

The observations required are more readily and less expensively made by sending up hydrogen-filled balloons carrying automatic weather-detection devices in an assembly known as a radio sonde.

Although the first British design for a radio sonde was laid down in 1938 by H. A. Thomas, the first British one to go into production was designed at Kew in 1940 by E. G. Dymond.

It is an instrument containing elements for the measurement of humidity, temperature and pressure, conveyed to ground receiving stations by means of a battery-operated transmitter. The instrument is attached to a rubber balloon of special strength and size which is released when required and allowed to freely ascend. A light-weight metal reflector attached between the balloon and the radio sonde enables the flight to be followed by radar, and a parachute allows a gentle journey earthwards once the balloon has burst on reaching its height limit.

While an ordinary coloured balloon filled with hydrogen at a funfair will reach something like 30,000 feet before bursting, the meteorologist needs heights far in excess of this, and so balloons are made which will attain heights of around 130,000 feet with a relatively inexpensive balloon, although the frequent use of such a device from the many weather stations in operation would provide a fairly formidable daily expense account.

In flight, while the ground radar measures change of speed and direction due to the influence of the winds encountered at different heights, the other instruments are at work recording temperature, pressure and humidity, and the transmitter is busy sending coded signals which are interpreted by the ground operator who is able to plot the points on a calibrated curve before him, from which is obtained the temperature and humidity at various pressures.

From all this information the weather experts can determine the condition of the upper air and predict its probable influence on the future conditions at the Earth's surface.

During this chapter we have touched upon the subject of humidity and have hinted at its importance in weather forecasting. The following chapter is devoted entirely to the effects of the water vapour in the atmosphere and the significance of its humidity.

3/THE CAUSE AND EFFECTS OF WATER VAPOUR

'Mountains cool the uplifted vapour,
converting it again into water'

Aristotle

The gaseous bubble of atmosphere in which our Earth is encased is composed of several dry gases, those in the greatest amount being oxygen and nitrogen, and all of them retaining a constant percentage at all times; but co-existing with them is a moist vapour which varies considerably in amount and which does not always remain in its gaseous state. This is water vapour, and it is always present in the atmosphere in varying amounts per quantity of air, perpetuating its existence by the process of condensation and evaporation, now being seen, now not being seen (Fig 4).

If there were no process of condensation and evaporation the amount of water vapour originally present in the atmosphere would long ago have been absorbed into the general nature of the atmosphere and would have remained in a state of equilibrium with its neighbours.

Vapour is transferred into the air by the process of evaporation from the seas and oceans, lakes and streams, and, in fact, anything and everything that at any time is in a state of wetness or dampness – even to the washing on the garden line, for how else would the water content of it disappear and allow the garments to become dry?

So the water from the surface of the Earth gently sheds minute and invisible particles of water vapour into the surrounding air, and as this vapour is rising only from the surface of the source in contact with the air, the area of the surfaces affects the quantity and speed of evaporation.

In addition to the effects of surface contact, the vapour content of the air is largely influenced by the temperature, since the atmosphere can only absorb a certain amount according to its temperature at any given time. Therefore, when a sample of air reaches the limit of the amount of vapour it can hold in

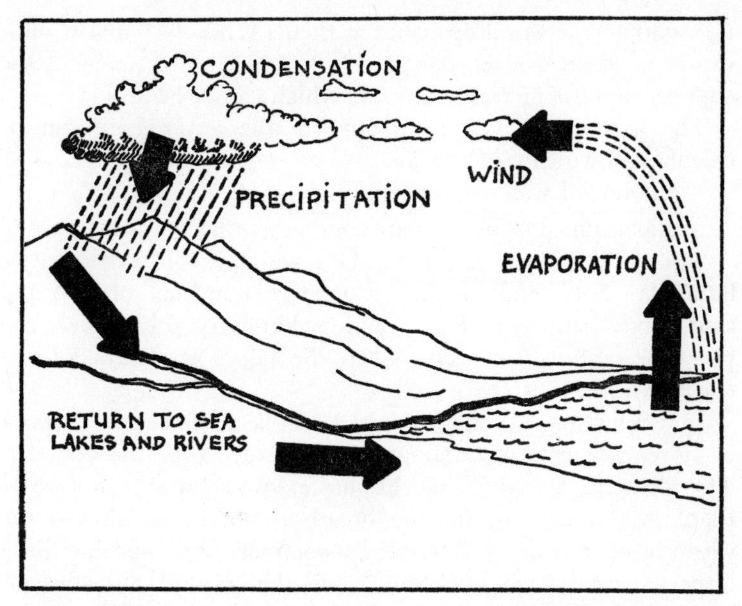

Fig. 4 *Cause and effect of water vapour.* Water evaporates from the ocean in the form of vapour. It is taken inland in this gaseous state by the wind, where it condenses and falls as snow or rain, when it begins its journey back to the sea in lakes and rivers.

invisible suspension, evaporation ceases and the air becomes saturated.

Until this time the water vapour in the air remains invisible, but when the rising saturated air becomes cooled below the level known as the dew point, it becomes visible through the process of condensation and the result we observe is in the form of clouds, for clouds appear when the rising air causing them has been cooled below its dew point.

It can be seen, then, that the water vapour content of the air becomes apparent only when changed into water droplets in the form of cloud or dew, this process being known as condensation. When the water droplets change from water to vapour again, the process is called evaporation.

Fog, for example, is normally caused by the condensation of water at ground level, or by smoke or dust being unable to rise, and the difference between fog and mist is in the size and transparency of the drops causing them. When walking in these clouds at surface level, one is able to feel the dampness of the copious supplies of water droplets which cause them.

The amount of vapour present in the atmosphere can be calculated by means of a ratio :

$$\frac{\text{amount of water vapour in unit quantity of air}}{\text{maximum possible vapour content at that temperature}}$$

The result is given as a percentage which is called the relative humidity. Saturated air has a relative humidity of 100 per cent. Fog occurs when the relative humidity rises above 100 per cent, and will disperse when the figure is returned to its correct value.

In winter, one's breath becomes visible as the water vapour in it reaches condensation point in contact with the cold air. This demonstrates that cold air has a lower capacity for water than does warm air, for the breath is not visible during the warmth of a summer's day. In summer, the morning haze disperses rapidly as the Sun heats the ground and the air becomes dry enough to absorb the evaporating water in invisible suspension.

Humidity

The humidity of the air is the degree of its approach to complete saturation, and between the vapour present in the air and the temperature of the air is the important link which enables us to measure the relative humidity. When the temperature is low, the amount of vapour capable of being absorbed is small, but when the temperature is high and the air is dry, it is capable of holding a greater proportion of water in invisible suspension.

The amount of water vapour present in the air is dependent upon its temperature and is called the relative humidity.

As the temperature falls, the relative humidity increases as the capacity of the air to absorb up to the maximum progressively becomes less. As the fall becomes persistent, the point of 100 per cent humidity, when the air is said to be saturated or at its dew point, is reached.

Further drop in the temperature renders it impossible for the air to support the weight of vapour already present, and the moisture clings to the miriads of dust particles to form fog, cloud or dew.

Formation of clouds

Clouds form as a result of the condensation of water vapour.

On sunny days the heat of the Sun causes the Earth to radiate heat waves, and some areas of the Earth react to this heating more slowly than others. Hard earth heats faster than soft earth, large areas of water take heat slower than areas of trees and bushes, but from all of these surfaces radiation will take place to some degree, depending upon the type of surface and the conditions causing it.

Hardly anyone can fail to notice how, on a sunny day, the surface of a roadway appears to be shimmering with rising trails of smoke.

This is the effect of radiation, and over large areas where this occurs, a sizeable portion of the air above it becomes heated and begins to rise, causing a sudden cool breeze as it does so. As the bubble of air rises, it expands and consequently begins to cool, while there is an immediate replacement of cool air below. This is brought about by the atmosphere trying to equalise the unbalanced state caused by the sudden breaking away of this area of hot air.

If this air rises high enough, it will reach the level where condensation takes place (the dew point) because of the lowered temperature, and as a result a cloud will begin to form, and there may be visible 'streets', or a series of clouds at the same level, above the column of rising air.

At other times, cloud is formed when an area of warm air is forced upwards over a layer of cold air. The highest clouds are formed when the water vapour, in rising, is turned into ice crystals.

Measurement of humidity

Hygrometers are instruments which measure the degree of dampness of the atmosphere. This is the amount of water vapour, expressed as a percentage, contained in the atmosphere compared with the maximum amount which it can hold at a given temperature, namely the dry air temperature at the time of recording.

The first hygrometer was developed by Leonardo da Vinci in the sixteenth century, an absorption type which was later further developed by Hooke and Saussure.

Some instruments measure the dew point of the atmosphere, while others, correctly known as psychrometers, show wet and dry bulb temperatures, and from these the relative humidity can be found by applying their readings to specially prepared tables.

The wet and dry bulb principle was developed in 1802 by Boeckmann.

Psychrometers are hygrometers of the evaporation type and are also known as wet and dry bulb thermometers. They consist of two identical thermometers mounted side by side, the bulb of one being covered with a muslin sock which trails into a small container of distilled water, so that the muslin is always damp; this is the wet bulb thermometer (Plate 7).

According to the dryness of the atmosphere, water evaporates from the damp muslin causing a cooling effect which drops the temperature of the thermometer, thus giving a lower reading than the thermometer without the sock. So, obviously, there is a difference between the two readings.

An air mass which is already damp when it reaches the locality of the thermometers will not cause evaporation to such a degree as dry air, and so the wet bulb reading will not be much lower than the dry bulb reading. The two may, in fact, give identical readings in saturated conditions. The cause of this is not so much that the wet bulb is not experiencing much evaporation, as that the dry bulb is probably covered with moisture which brings its reading in line with its companion. In effect, they are both acting as wet bulb thermometers at that particular time.

On a hot day, the rate of evaporation will be rapid and the readings will be of considerable difference.

Using the temperatures given by both thermometers we can, by applying the result to prepared tables (see facing page) or to a special slide rule calculator, find the relative humidity.

If the air temperature is below 32°, some caution is required

Table for finding Relative Humidity (per cent)
Depression of Wet Bulb

Dry Bulb °F	0°	1°	2°	3°	4°	5°	6°	7°	8°	9°	10°	11°	12°
90	100	96	92	88	84	81	77	74	70	67	63	60	57
88	100	96	92	88	84	80	77	73	69	66	63	59	56
86	100	96	92	88	84	80	76	72	69	65	62	58	55
84	100	96	92	87	83	79	76	72	68	64	61	57	54
82	100	96	91	87	83	79	75	71	67	64	60	57	53
80	100	96	91	87	83	79	74	70	66	63	59	55	52
78	100	95	91	86	82	78	74	70	66	62	58	54	50
76	100	95	91	86	82	78	73	69	65	61	57	53	49
74	100	95	90	86	81	77	72	68	64	60	56	52	48
72	100	95	90	85	80	76	71	67	63	58	54	50	46
70	100	95	90	85	80	75	71	66	62	57	53	49	44
68	100	95	90	84	79	75	70	65	60	56	51	47	43
66	100	95	89	84	79	74	69	64	59	54	50	45	41
64	100	94	89	83	78	73	68	63	58	53	48	43	39
62	100	94	88	83	77	72	67	61	56	51	46	41	37
60	100	94	88	82	77	71	65	60	55	50	44	39	34
58	100	94	88	82	76	70	64	59	53	48	42	37	31
56	100	94	87	81	75	69	63	57	51	46	40	35	29
54	100	93	87	80	74	68	61	55	49	43	38	32	26
52	100	93	86	79	73	66	60	54	47	41	35	29	23
50	100	93	86	79	72	65	59	52	45	38	32	26	20
48	100	92	85	77	70	63	56	49	42	36	29	22	16
46	100	92	84	77	69	62	54	47	40	33	26	19	–
44	100	92	84	75	68	60	52	45	37	29	22	16	–
42	100	91	83	74	66	58	50	42	34	26	18	–	–
40	100	91	82	73	65	56	47	39	30	25	–	–	–
38	100	91	81	72	63	54	44	39	31	22	–	–	–
36	100	90	80	70	60	54	44	35	26	18	–	–	–
34	100	90	79	70	60	50	41	31	21	–	–	–	–
32	100	89	79	68	57	47	36	27	17	–	–	–	–
30	100	88	76	65	53	43	33	22	–	–	–	–	–

in reading the wet bulb thermometer. If the wet bulb reading is higher than the dry bulb reading, it cannot be recorded as the instrument is not in working order and should be cleaned and the water bottle replenished with distilled water or rain water.

Since evaporation takes place from ice as well as from water, an observation taken when the muslin is frozen may be acceptable.

After every observation in frosty weather it is necessary to immerse the muslin and bulb in water so that by the time of the next reading there will be sufficient ice to provide an acceptable reading.

Once the temperature has risen above 32° it is necessary to

remove any ice which adheres to the muslin; a few seconds immersion in warm water is therefore needed before returning the muslin to its bottle of water, and a time lapse must be allowed before the next reading so that the thermometer has been subjected to a proper amount of cooling to get it back to normal conditions.

Professional and meticulous amateurs use a louvered box known as a Stevenson screen in which to keep their thermometers, and each thermometer is individually mounted, so that the dry bulb and wet bulb types are mounted vertically side by side, and they share the screen with a thermometer which is mounted horizontally and is made to indicate the maximum day temperature, and a horizontally mounted alcohol thermometer which is made to indicate the lowest temperatures. In other words, the maximum and minimum thermometers introduced in Chapter II.

Whilst on the subject, it is as well to establish that evaporation is the physical process by which liquid water becomes water vapour, and the instrument used to record the degree of evaporation is known as an evaporation gauge or evaporimeter. In simple form, the gauge consists of a glass tube about 9 inches long and containing water, the level of which can be read off against a scale measured in centimetres. The water in the tube evaporates by way of a disc of porous paper held in place by a clip, the degree of evaporation being evident by the alteration of the level of the water in the tube.

Recording hygrometers or hygrographs

The hygrometer provides a continuous record of the relative humidity of the atmosphere, the hygroscopic element being a human hair so treated as to remove all traces of grease.

As the humidity increases, the hair lengthens and affects the movements of simple mechanisms of which a recording pen is part. This pen is caused to trace a line on a specially prepared graph.

Thermo-hygrographs

This instrument records temperature as well as humidity by using two pens marking the same chart. In order to prevent the pens jamming each other, the recording pen for humidity is set 4 hours behind that used for recording the temperature.

It is, of course, very convenient to have all the time-absorbing work taken out of daily observations by automatic recording instruments, but a brief visit to the retailer of such instruments will soon show that £80 and £90 per unit is not uncommon. It is therefore an expensive hobby when taken to its sophisticated conclusion. Fortunately, the local weather forecaster can make all the necessary observations by means of relatively inexpensive equipment, although he would be advised to purchase the best he can reasonably afford for the sake of accuracy and durability.

Apart from this, it is possible to make a few simple calculations for oneself.

Calculation of the dew point

It has been said that condensation will take place when the dew point has been exceeded. Meteorological stations are able to give the dew point for all levels of the atmosphere, having arranged for readings to be taken either from aeroplanes on routine 'met' flights or by special balloon, but whatever the altitude of the readings, the system is the same and, failing a more accurate means, the following method provides a rough figure on which to work.

Take the difference between the wet and dry bulb temperatures and subtract it from the wet bulb reading:
Example: dry reading=50°F; wet reading=48°F; difference is 2°F. Subtract thus: 48−2+46°F, which is the dew point.

When the temperature reaches this figure, condensation will occur and cloud or dew may be expected to form. The following table will give the dew point over a wide range of readings.

Table for finding the Dew Point (°F)

Dry Bulb °F	Depression of Wet Bulb												
	0°	1°	2°	3°	4°	5°	6°	7°	8°	9°	10°	11°	12°
90	90	89	87	86	85	83	82	80	79	77	76	74	73
88	88	87	85	84	83	81	80	78	77	75	74	72	70
86	86	85	83	82	80	79	78	76	75	73	71	70	68
84	84	83	81	80	78	77	75	74	72	71	69	67	66
82	82	81	79	78	76	75	73	72	70	68	67	65	63
80	80	79	77	76	74	73	71	69	68	66	64	62	61
78	78	77	75	74	72	71	69	67	66	64	62	60	58
76	76	75	73	72	70	68	67	65	63	61	60	58	55
74	74	72	71	69	68	66	64	63	61	59	57	55	53
72	72	71	69	67	66	64	62	61	59	57	55	52	50
70	70	69	67	65	63	62	60	58	56	54	52	50	47
68	68	66	65	63	61	60	58	56	54	52	49	47	45
66	66	64	63	61	59	57	56	53	51	49	47	44	42
64	64	62	61	59	57	55	53	51	49	47	44	41	38
62	62	60	59	57	55	53	51	49	46	44	41	38	35
60	60	58	56	55	53	51	48	46	44	41	38	35	32
58	58	56	54	52	50	48	46	43	41	38	35	32	28
56	56	54	52	50	48	46	43	41	38	35	32	29	25
54	54	52	50	48	46	43	41	38	35	32	29	25	20
52	52	50	48	46	43	41	38	36	32	29	25	20	16
50	50	48	46	43	41	39	36	33	29	25	21	16	10
48	48	46	44	41	39	36	33	30	26	22	17	12	4
46	46	44	42	39	36	34	30	27	23	19	13	6	–
44	44	42	39	37	34	31	28	23	19	15	8	–	–
42	42	40	37	34	32	28	25	20	16	9	–	–	–
40	40	38	35	32	29	26	22	17	11	8	–	–	–
38	38	35	33	30	26	22	18	15	10	3	–	–	–
36	36	33	30	27	23	21	16	11	5	–	–	–	–
34	34	31	28	25	22	17	13	7	–	–	–	–	–
32	32	29	26	22	19	14	8	–	–	–	–	–	–
30	30	27	23	20	15	10	4	–	–	–	–	–	–

Calculation of the condensation level

Another useful calculation for the amateur is that which enables him to determine the height at which cloud may be expected to form once he has established the dew point.

Stratus cloud is probably the most inconvenient of the clouds since it may form rapidly over high ground and saturate the wayfarer, and at higher level may impede the vision of pilots.

Using the method described for finding the dew point, it is possible to make rough calculations of the approximate height to expect cloud.

If the observer has reason to suspect near-saturated air, he may proceed thus:

a. Observe the dry and wet bulb temperatures.

b. Find the difference between them.

c. Subtract the difference from the wet bulb reading, thus:

Example: dry reading =65°F; wet reading=61°F; difference =4°F. Subtract this from wet bulb temperature, so: 61−4 =57°F, this being the dew point and telling us at what temperature condensation will occur.

For the final step in finding the height of condensation when the wind gains enough force to raise the air high enough, we proceed as follows:

a. Take the number of degrees the dew point is below the dry bulb temperature,

b. multiply this figure 220 times.

Example: the answer to instruction a. is 8°F; the answer to instruction b. is 220×8=1760 feet.

This is the probable height of the condensation level from the point of observation.

If we are in an air mass in which the rising air cools at the rate of 3°F per 1000 feet, and the dew point is 3°F lower then the dry bulb temperature, the air must rise to 1000 feet before its water vapour content will condense out as cloud. Similarly, in a sample of air in which the rate of cooling is 5·4°F per 1000 feet, and the dew point is 5·4°F below the dry bulb temperature, that sample of air must rise to 1000 feet to experience condensation.

When the stratus forms upon the hills and the barometer falls against a rising temperature, rain may be a safe forecast.

Morning stratus in summer months may precede a fine outlook and will evaporate as the Sun heats up the ground.

Dew

Dew is formed by the same process of evaporation and condensation that is responsible for cloud, rain, fog and mist, and assists in providing vegetation with the moisture it requires to maintain healthy plants.

Certain daily conditions point to the possibility of dew forming. The daylight hours must be fine and warm, twilight and night conditions must be cool with no air movement and no clouds.

A pocket mirror set about an inch above the ground will prove that at this time there is an amount of dampness at the surface.

Vegetation, characteristically losing heat rapidly at night, becomes cold and the water vapour in the air condenses on the grass and leaves of plants.

The pocket mirror will advise of this formation at once as the invisible vapour condenses on to the cold glass, producing miriads of minute water droplets. This occurs in much the same way as it would if you were to enter a warm room carrying a glass of ice-cold water. The air in the room, upon contact with the cold glass, condenses and forms a mist on it.

An interesting experiment, which will support the connection between radiation and the formation of dew, can be made on an evening when a heavy dew is expected. On this night stretch a square of muslin on 4 canes so that it covers an area of grass about 4 feet square and about 10 inches high. Later, when the dew has formed on everything else, or perhaps in the morning before evaporation has begun, check the grass beneath the muslin awning and you will find that there is hardly any dew deposited on it, and yet there is dew all around. The muslin has formed a kind of blanket above the grass and has prevented the radiation of heat waves from the grass dispersing rapidly into the air, so that the grass was not chilled enough to attract the water droplets from the atmosphere.

For a similar reason there will often be little or no dew

deposited during a cloudy night, because the clouds return heat to the ground so that it does not become chilled enough to draw water from the air. But after a hot, dry day, with a cloudless evening and no rain to refresh the countryside, the leaves of plants and trees are cooled and draw the vapours from the air to drink.

Conversely, some damaging conditions may be brought about at the end of a fine spring day in this way. The sky will be cleared and the air perfectly still, and the thermometer will rapidly fall.

The cold air in sinking to the ground freezes the condensation at that level, causing frost.

Dew is produced in serene weather and in calm places

Aristotle

Dew is an indication of fine weather; so is fog

Fitzroy

Frost

The spring frost is the most unkind to our crops and is the one to be watched and guarded against. Black frost occurs when the air conditions are not favourable for the formation of dew, and the temperature falls below the freezing point of water and freezes the condensation. Hoar frost is formed by the dew being deposited as minute ice crystals when the night temperature falls below freezing.

In winter and spring, when the wet bulb is the lower, there is likely to be frost, especially if these readings are taken about 6 p.m. Further caution should be indicated by winds from the north-west, north-east and east.

In spring, frost may follow upon a warm day when the barometer reads high, all clouds have dispersed, and the air is calm and dry; then frost is probably on the way.

On the other hand, if there is damp air with high cloud amount, the blanket so formed will not allow the warmth to escape from the ground, and frost is unlikely. Breeze and wind will also prevent the formation of frost, as will high humidity with a moderate breeze.

Our severest frosts are experienced during an area of high pressure over northern Europe, when the anticyclone from

Siberia comes further west than is normally expected. This movement opens up a corridor in the weather systems which allows Arctic or icy Continental air to overtake large areas of the British Isles.

Radiation frosts, occurring during the March-May period, are particularly important to forecasters since the blossom and fruit buds can be killed off so easily at that time. Ground frosts have been reported as late as the beginning of June.

If the frost occurs soon after rain, which means that the countryside is covered by a film of water drops, the damage is more severe than when it follows a dry night.

If the temperature drops rapidly after sunset and the morning thaw comes suddenly, the damage is more severe than it would have been had the change been slow and regular.

Aircraft vapour trails

Before cloud particles begin to form the air must become saturated, or almost so, either by the reduction of temperature or by the addition of more water vapour. In nature, the physical conditions for such formations are produced by the characteristics of the air mass in which they occur, but artificially manufactured cloud is a man-made product with which we are readily familiar. Examples are the appearance of the saturated cloud of steam from a boiling kettle, or the visible exhaust from a car in cold weather; but the most beautiful and the most observed man-made cloud is that caused by high-flying aircraft.

For each gallon of fuel consumed by the piston engined type, nearly $1\frac{1}{2}$ gallons of water vapour are expelled into the surrounding air, and as the temperature is well below zero, probably between -15° to -45°, the vapour condenses into cloud in just the same way as when water vapour is forced up to its condensation level to form the same type of cloud. The cloud so formed is called cirrus, and it occurs in temperate latitudes at heights in excess of 22,000 feet in winter, and over 28,000 feet in summer. As we might suppose after a little thought, it may occur at considerably lower levels in exceptionally cold climates.

The condensation trails, or contrails as they are often called, are to be seen crossing the sky during all seasons, for at that height the conditions for making cirrus cloud are always freezing, no matter what the season is on Earth, and for that reason the cirrus clouds are composed entirely of frozen vapour particles, appearing fibrous and feathery, stretching in long lines across the blue sky. Generally, the trails disperse and die out, but sometimes they are absorbed by the natural cirrus already in existence.

Conditions of the air

From the few facts we have so far discussed, we may safely assume that air does not remain in one state or condition for very long and that, as a result of this change of state, it may be classed as being variable in its water content. By way of a summary, we may itemise the following conditions:

a. Dry air is a mixture of nitrogen and oxygen and other gases, with a variable amount of water vapour. This is the basic construction of our atmosphere.
b. Humid air is air with an amount of water vapour below its saturation point.
c. Saturated air is air containing water vapour at saturation point.
d. Wet air is air containing saturated vapour and rain, snow or ice.
e. Super-saturated air is air with a percentage of water vapour greater than saturation vapour pressure.
f. Super-cooled air is air containing saturated vapour and super-cooled liquid vapour.

Ice

Owing to the movement of the molecules forming it, water expands when heated and, conversely, owing to the deceleration of the molecules, it contracts when cooled, and this contraction takes place until the temperature of the water is lowered to about 39.2°F. After this point the water expands until it freezes at its freezing point of 32°F. It is this expansion which makes an ice mass less dense than a similar mass of water, which is the reason that icebergs float, as do layers of ice on rivers and lakes.

From this it will be clear what when water freezes in the water pipe system of the house, the ice so formed will expand beyond the normal capacity of the pipe, and the pipe will split.

The damage is not at once apparent as the contents of the pipe are frozen, but when the thaw begins and the ice returns to its liquid state, the water will naturally escape through the split section.

It is this delay between the actual freezing of the water and the splitting of the pipe and the thaw which leads many people to believe that the bursting of the pipe occurs actually at the time of, and as a result of, the thaw.

Ice on aircraft

One of the hazards of flying is the formation of ice about the body and wings of the aircraft. The formation of hoar frost, rime and glazed ice are familiar occurrences on the ground, but during winter they are more frequently experienced in flight because of the lower temperatures of the upper air, and dangerous conditions are easily reached if due caution is not exercised.

When icing occurs on the leading edge of the wings, the aerodynamic characteristics of the aerofoil begin to change, the effect of which is to increase the air resistance and the stalling speed. Severe conditions may reduce cruising speeds to about 30 per cent of normal.

As the air resistance, or drag, as it is called, increases due to the excessive icing, it is necessary to increase engine power in order to avoid loss of lift, but in piston engined aircraft, it is possible for ice to form in the carburetter or on the airscrew, so that the extra power required is not available. In these conditions in a light aircraft the pilot looks at his indicators for loss of engine revs and a decrease in the air speed indicator.

Excessive and persistent icing on the airscrew destroys the original balance of the blades, and the possibility of chunks of ice being flung against the cockpit and fuselage is an added danger.

Additionally, the conditions which give rise to icing are likely to cause jamming of the control surfaces such as the ailerons, elevators and rudder, and ice may melt and break off from other parts of the aircraft and cause damage in being hurled backwards in the turbulence of the slipstream.

Modern airliners, although equipped to deal with icing conditions by means of de-icing arrangements, are, even so, likely to be subjected to dangerous conditions at critical altitudes.

The definition of the freezing level is the height above sea level at which a temperature of 32°F (0°C) is encountered. This height obviously varies according to seasons and general conditions, but the levels at which icing may occur over the British Isles are from about 10,000 feet in summer, down to about 1800 feet in winter.

Visibility

Our range of vision along the ground varies almost from day to day, and often from one part of the day to another, as atmospheric conditions change. Conditions of fog, haze or rain reduce the distance one can see with the naked eye in a horizontal line without obstructions; this ability to see is known as visibility.

Visibility is measured in the number of yards of horizontal vision, and the standard is laid down by the International Visibility Code, which is as follows:

Code No.	*Objects visible*		*Description*
	at	*not at*	
0	–	55 yds	Dense fog
1	55 yds	220 yds	Thick fog
2	220 yds	550 yds	Fog
3	550 yds	1100 yds	Moderate fog
4	1100 yds	1¼ miles	Mist or haze
5	1¼ miles	2½ miles	Poor visibility
6	2½ miles	6¼ miles	Moderate visibility
7	6¼ miles	12½ miles	Good visibility
8	12½ miles	31 miles	Very good visibility
9	31 miles or more	–	Excellent visibility

For night observations, lights at measured distances and of known intensity are used as a means of estimation.

Around and about the country, from farm to farm and from coast to coast, one hears many sayings concerning visibility and the weather, most often about the possibility of rain being imminent when distant objects are seen with greater clarity than usual; for example, Brighton, Worthing and Bournemouth observe that there will be rain in a day or two when the Isle of Wight is perfectly visible, since when the weather is fine the Isle is almost invisible from these points, while from Cornwall we learn that:

'When the Lizard is clear, rain is near.'

Sound and the weather

In much the same way as clear visibility is said to herald rain, sound appears to be affected by similar atmospheric changes, for we are told in weather lore that 'A good hearing day is a sign of wet', and that 'Sound travelling far and wide, a stormy day will betide'. The observation is also to be found recorded by Bacon as follows: 'The ringing of bells is heard at a greater distance before rain; but before wind it is heard more unequally . . .'

Locomotive and factory whistles may be heard with extra clarity during the changing conditions before a storm, and when the vibration of bells appears to fluctuate it is a sign of wind accompanying the rain.

Precipitation

Precipitation is the collective term for all moisture which condenses into the atmosphere, and therefore includes snow, hail, sleet, fog and rain, all of which can be measured by means of a rain gauge, which we shall discuss shortly.

Rain and drizzle

'The rain it raineth every day'

Shakespeare

A cloud is formed of small water droplets supported on rising air and maintained within the turbulence of the cloud, so that we may say that basically the cause of rain is the ascent of air into the higher regions of the atmosphere, whether it be from the effect of mountains causing the air to rise, moist air currents being forced upwards by colder, drier air beneath them, or moist air being drawn upwards over the area of low pressure at the centre of a depression.

The direction from which the wind comes is also an important factor influencing the rainfall, since it may arrive from the sea area and will therefore have a high moisture content, much higher than a wind which has come across wide areas of land and is comparatively dry.

Different winds may meet and mix, thus combining their water content which can result in precipitation.

Any circumstance which lowers the temperature of the air may cause rain from the appropriate cloud formations because, as the size of the water drops within them increases and becomes too heavy to be supported as cloud, they fall as precipitation.

If the precipitation is very fine and appears to float down, it is termed *drizzle,* whereas the larger type, the drops of which may measure up to 5.5 millimetres in diameter, is called *rain.*

Drizzle is associated with the thin layer type of cloud, and rain is associated with the heaped cumulonimbus or nimbostratus.

There is also the rate of fall to be considered, for as we know, we get heavy rain and light rain, some falling slowly, and some falling quickly, the larger drops coming with greater force than the smaller ones. The speed of all precipitation is affected by the resistance of the air itself, and there comes a point where the resistance is equal to the weight of the drops, at which time the speed of descent is reduced to a constant rate of fall known as the 'terminal velocity'. In approximate figures, fine drops fall at 2 miles an hour, medium drops at 15 miles an hour, and large drops at 18 miles an hour.

There are, of course, varying degrees of rain intensity, and these are briefly outlined in the following manner: slight rain is rain of low intensity which may be composed of numerous small drops or scattered large drops; moderate rain is rain which falls fast and forms puddles; while heavy rain is of the type that crashes down during a thunderstorm.

As distinct from observations, showers are caused by vigorous convective processes which increase their intensity above other forms of rain.

This kind of precipitation, rain and showers, is gauged by its ability to accumulate in the standard rain gauge at a certain rate, so that we may devise the following simple table:

Intensity	*Rain*
slight	not more than about 0.5 mm (0.02 in) per hour
moderate	between 0.5 and 4 mm (0.02 to 0.16 in) per hour
heavy	more than about 4 mm (0.16 in) per hour

Intensity	*Showers*
slight	less than about 2 mm (0.08 in) per hour
moderate	about 2 to 10 mm (0.08 to 0.04 in) per hour
heavy	about 10 to 50 mm (0.4 to 2 in) per hour
violent	more than about 50 mm (2 in) per hour.

Over large areas of sea and ocean, the air is not subject to extremes of temperature since the water is slow to warm up and slow to cool by comparison with large land areas. Therefore, the rainfall is less over the sea than over the land. Naturally, the inland weather is tempered by the prevailing winds, which means that we in the British Isles receive the moist, mild Atlantic air coming in on the prevailing south-west winds and that the south-western extremities of the country enjoy the best and the worst of the weather as it arrives from that quarter, with the highest rainfall centred in Ireland.

In the Highlands, the Lake District and the Welsh mountains the annual rainfall exceeds 125 inches, while the eastern coast is comparatively dry with a mean average of 20–25 inches; we can quote two extremes with The Stye, Cumberland at about 175 inches, and Dagenham, Essex at about 19.3 inches.

The annual rainfall is the total number of inches of rain

which has fallen during the year at any one place, and a calculation of the average fall over several years is taken as the mean annual rainfall.

Extreme examples of wet and dry regions are shown by certain places in the Gobi and Sahara deserts where rain never falls owing to the constant nature of the temperature maintaining what little vapour there is in a state of suspension in the dry, hot air; and by the Khasi Hills in Assam, south of the Himalayas where the annual average rainfall is 428 inches.

Nevertheless, without rain, the world would be an enormous desert with no life and no vegetation, and so rain, and the forecasting of it, is of great importance, especially as all of our water supply must originally come from rain.

Abnormal rainfall in our changeable climate sometimes causes unexpected havoc with life and property, and we often blame the latest scientific device for its origin. The detonation of bombs and other forms of violent explosion has been said to be the cause of storms and unseasonable rainfall, and even this would appear more probable than a report which appeared in the *Evening News* of 1903, which read:

'In trying to account for the extraordinary vagaries of our climate this year explanations have taken a wide range, and the latest theory to account for the continued wet is that the wireless telegraphy operations of Signor Marconi are responsible.'

Snow and sleet

'He casteth forth his ice like morsels
He giveth snow like wool'

Psalm cxivii, 17 and 16

When water vapour condenses in air temperatures below freezing point, it forms minute ice particles and these particles unite, so forming large ice crystals called snow-flakes; in other words, it is the frozen moisture which falls from the clouds when the temperature is 32°F or lower.

When the flakes become too heavy to be supported in the cloud manufacturing them, they fall towards the ground. If, in falling, they pass through temperatures above freezing point, they become a mixture of unmelted flakes and partly melted flakes and rain, which we call sleet. If on the way down the temperature is such that it causes all of the flakes to melt, the precipitation reaches us as rain.

Snow is associated with the great cumulo-type clouds, but may be produced from both heap or layer cloud at the correct temperature.

Each snow-flake is composed of many snow crystals; upon minute examination, it will be found that each crystal is six-sided, and of all the variety of beautiful patterns to be found, no two are identical.

Snow-flakes are composed of these tiny ice crystals held together in a network formation which allows for more air than snow in the whole flake. This accounts for the lightness of freshly fallen snow.

A good snow fall covers the Earth like a feathery blanket and acts as a protection to plants during freezing weather.

Since snow is a poor conductor of heat, it keeps the warmth of the Earth in and prevents rapid radiation, so that the lower layers of snow are comparatively warm while the surface layers can be exceptionally cold and even under blizzard conditions.

The easterly winds from the Continent bring us the coldest weather, usually between January and March, so that snow is more likely and, indeed, more frequently recorded in late winter and early spring, while, in spite of traditional sentiments, it rarely falls during Christmas.

Snow will fall to the ground and 'lay' if the temperature and general conditions are sympathetically inclined for maintaining it in its flake-like form.

It can, over a long period, become compressed and freeze into hard ice, or if conditions stay cold, calm and dry with clear skies, it will remain crisp, and may become hard, even if temperatures rise to 37°–38°F.

Generally, with the usual conditions which set in after snowfall, the snow will melt at 34°–35°F. This process is called thaw.

Hail

> 'And Moses stretched forth his rod toward heaven; and the Lord sent thunder and hail, and the fire ran along the ground; and the Lord rained hail upon the land of Egypt.'
>
> *Exodus 10, 23*

When raindrops freeze in the regions of low temperatures, they form into transparent units of ice and, because the cloud in which they originate cannot support their weight, they fall towards the ground. If on the way they pass through higher temperatures, they may melt into rain, or into smaller particles which are carried upwards on air currents back into the regions of freezing conditions, where they again turn into hard ice.

On the way down once again, they collect a coating of water which freezes on contact and forms a shell around the first ice pellet. Another strong up-current may hurl the ice back into the intense cold where another shell of ice forms around the second and increases the size and weight of the pellet. This process may continue many times, with the enlarging pellet dashing towards the Earth, then being caught in strong up-currents to be whisked back into the cold, gaining another shell and falling back in its endeavour to comply with the gravitational forces of the Earth.

Finally, at some time during this process, the pellet of ice becomes too large and too heavy to be supported on the up-current, and falls to the Earth, often at very high speed, threatening to smash windows and stinging the face and hands.

This is called hail and may consist either of small pellets of clear ice which is frozen rain, or of puffs of white ice, or of the large-size ice particles which we usually call hail-stones.

During the next heavy storm, try to find a large hail-stone and, before it melts, cut it in two with a sharp carving knife and examine the interior under a magnifying glass. It will be seen that it is composed of several layers of alternating texture – rather like looking at the growth rings on a sliced tree trunk.

There is a layer of clear ice, which represents frozen water vapour, and there is a layer of white ice, which represents frozen rain mixed with air, so that if we are able to count the number of these layers, or growth marks, we can determine the number of times that piece of hail has ascended and descended in the atmosphere before coming to Earth. Having caught one's hail-stone, it is best to perform the dissection in the coldest possible conditions to prevent melting before its make-up has been observed. Hail is associated with the cumulonimbus cloud.

Showers

> 'When ye see a cloud rise out of the West, straightway ye say, There cometh a shower; and so it is.'
>
> *Luke xii, 54*

Showers of rain, snow or hail, are associated largely with the cumulus or cumulonimbus cloud. When precipitation falls for periods of short duration with breaks of fair to good weather between each fall, it is termed a shower.

It may be of interest, in passing, to note that the beautiful hymn 'Rock of Ages' was inspired by tempestuous rain having driven a traveller, the Reverend Toplady, to shelter in a deep cleft in the Cheddar Gorge.

> 'Rock of ages, cleft for me,
> Let me hide myself in Thee.'

Warm rain

Although it can also occur over land areas, warm rain is more especially associated with cumulus over the sea, and appears more commonly in warmer climates and during the English summer.

It is observed as hazy vertical columns joining the cloud base to the horizon, the columns composed of falling warm rain.

Whereas the turbulence within land cumulus hurls the cloud's water droplets into the uppermost sections of the cloud, where they may either evaporate or become absorbed by the frozen higher regions, the less vigorous thermals from the sea keep the water droplets at a low level and so saturate the air until rain begins to fall.

Fog

The precipitation responsible for fog, which reduces visibility, is caused by the condensation of water vapour forming a cloud at ground level when the air is cooled below its dew point by contact with colder ground. The mixing, or spreading, of fog may be brought about by a light wind.

When the night is clear and the wind is light, the air near the ground cools rapidly by radiation below its afternoon dew point, and condensation takes place. Damp air and damp ground are favourable conditions for the appearance of fog.

Fog may be widespread and in winter can last for several days with little variation, or it may be local and confined to valleys and depressions in the ground, leaving higher areas with normal clear visibility.

In warmer weather, fog will usually be cleared as the morning sunshine warms the air and drops the relative humidity, but failing this the fog will prevail until a wind blows it away.

Long periods of fog are often due to a mass of warmer air having been forced to rise, so that the temperature in the fog area is fairly low and that of the air mass above it may be showing an increase. This is known as a temperature inversion, for the lapse rate from ground level does not fall continuously with increase of height, meaning that the air is warmer at about 1000 feet that it is below.

Eventually, this warm air mixes with the fog and evaporates it away.

In winter, fog may result from warm air moving in over cold ground or cold sea, such as when, during a cold spell, a mass of humid air arrives from the south-west and produces a blanket of persistent fog.

Fog at sea is formed mainly by air currents moving from a warm sea over one of a cooler nature. In this case, the lower layers of the warmer air are cooled by the surface of the sea, and as the temperature is brought below its dew point, the fog will form.

The depth of a fog will depend upon the length of time it has persisted. Some winter fogs, having built up over a few days, reach 500 to 1000 feet above the surface.

Fog caused by cooling at night usually has a depth of about 50 to 100 feet, while sea fog may be shallow enough to envelope a ship's superstructure, leaving the mast tops in sight, or to obscure the lower regions of a lighthouse while the light continues to flash its signals through clear air.

Fog and mist are both due to suspended copious supplies of water droplets in the air, and are defined only by the difference in their transparency, thus: visibility 1100 yds to $1\frac{1}{4}$ miles = mist, visibility 1100 yards to nil = fog.

There are three causes of fog, namely radiation, movement and mixing. Radiation fog is caused by the ground radiating its heat into a clear, crisp night sky and cooling the surrounding air below its dew point, and is most frequent in autumn and winter. Movement fog is caused by a mass of warm air moving over a cooler surface and itself becoming cool enough to reach dew point, an effect experienced at sea. Mixing fog is caused by moist air mixing with cold, dry air, so reducing the temperature below its dew point.

A fog with a rising barometer may last several days, while a low barometer may mean that the fog will be dispersed by rain.

The rain gauge

While it is the function of the hygrometer to determine the amount of water vapour present in a mass of air, it is the function of the rain gauge to measure the amount of water when it condenses out.

Rain gauges were used in Korea as far back as 1442, but the first to be devised in England was by Sir Christopher Wren in 1662, although the first person to actually record rainfall was Townley of Burnley in 1677. The first self-recording rain gauge was developed by Crossley in about 1829.

The standard Snowden rain gauge consists of an outer container of copper, and is situated in an open place away from dripping leaves and buildings, with the rim set 1 foot above the ground (Plate 8).

Inside the top of the container, about 3 inches down, is a 5-inch funnel for collecting the precipitation and passing it into a collecting vessel beneath. The content of the container is poured into a glass measure and the amount present is read from a scale marked on its side.

If there is barely enough moisture to register, it is entered in the weather log as 'trace'.

The student is often disappointed by the small amount of water in his gauge, but he must consider that about ½ inch of rain represents over 50 tons of water per acre and, to break it down still further, that 1 inch of rain registered over an acre will represent a fall of some 22,624 gallons of water or, by weight, 101 tons per acre.

Snow cannot be measured until it has melted into water, but

because of its bulk compared with its condensation value it takes about 3 inches of clean snow to produce about $\frac{1}{4}$ inch of water.

All the forms of moisture that can be measured are grouped together, and the result is called the 'mean annual rainfall'.

Rainfall is greatest in the tropics and least at the Poles, while in those places not situated in the tropics the rainfall is greatest in mountain, hilly and coastal areas.

Colour indications in the sky

> 'Men judge by the complection of the sky
> The state and inclunation of the day.'
>
> *Shakespeare*

> 'Red clouds at sunrise foretell wind: at sunset, a fine day for the morrow.'
>
> *Bacon*

Of the various colours to be seen in the sky, green seems to provide the more sure means of forecasting, since the appearance of green or of a greenish hue above the sunset will be followed by rain during the succeeding 24 hours.

This presence of green is not confined solely to times of sunset. The tint may appear at any time during the day, although it would seem to be more readily recognisable over sea areas than over land, and varies in intensity.

In some parts of the world, observers say that it takes a lifetime to see green in the sky, and therefore renders the method void as far as they are concerned. Others however, see green frequently, and almost everyone in the British Isles must be familiar with the colour as it spreads across the threatening sky.

A clear blue or sea-green colour close to the horizon precedes showers in summer and snow in winter, while yellow is prognostic of high winds; and if the yellow blends into an unpleasant green it may indicate stormy weather and heavy rain.

Out in space, the Sun is actually white hot and would shed these white rays at us continuously if it were not for the modifying effect of the atmosphere between us. For example, as the Sun approaches the horizon its rays appear to be a brilliant orange. This is brought about by the fact that the Sun's rays are filtered through the thickness of the atmosphere, and the blue light is filtered out, leaving only the red and yellow rays to reach us. If there is a high percentage of dust or a prevailing haze between us and the Sun, its disc appears red instead of orange.

The most beautiful sky colouring appears at sunset, especially when there is a great variety of high clouds illuminated with mauve, violet, yellow, pink and red.

The reason for the difference in appearance of the Sun's colour as it moves in relation to the horizon, and for the different colours in the surrounding sky, is the varying thickness of water vapour which lies between the observer and the observed. At sunset, when the clouds appear at various heights and in different parts of the sky, so that various thicknesses of vapour are suspended between them and the Sun, they are tinted with different colours which are the residue of the original light after it has experienced filtering and refraction by the vapour content.

On a fine day, the dawn clouds appear tinted with red. As the Sun rises, the yellow component of its light is no longer absorbed and the clouds are turned to orange, and then yellow, and then white. At sunset, its rays fall progressively more obliquely on the clouds, and the colour cycle runs through yellow, orange and red.

If the morning clouds are red and menacing it is a sign of rain and bad weather, for the red colouring is produced by an unusual amount of vapour which filters out the blue rays and allows the red and yellow rays to pass through.

A soft, light blue sky is likely to produce settled weather, but dark blue, sharply outlining the clouds, is an indication of stormy weather; while copper tints around the cloud edges are caused by violent electrical disturbance and are the fore-runner of thunderstorms.

Mainly in the winter months, the easterly winds send us puffs of cloud which are heavily tinted with violet, and which accompany prolonged periods of rainless but cloudy, cold weather.

The colours in the sky are useful indicators to the weather and reward careful study. In fact, a great deal of specialised research has been done on this subject in the past in separating the various colour bands which appear in the atmosphere and measuring their intensity and width.

> 'Red sky in the morning,
> Shepherds take warning;
> Red sky at night,
> Is a shepherd's delight.'

The Spectroscope

The spectroscope is an instrument which focuses a beam of light through a prism and separates it out into the various colours which comprise white light. These colours are produced by the different wavelengths causing them and are seen to be red, yellow, green, blue and violet, although where the colours overlap and merge into each other there are intermediate hues such as orange, yellow-green and blue-green.

It follows that as these various bands of light arrive at our spectroscope from such great distances across space, they must, in some way, be affected by the variations of the invisible atmosphere through which they pass.

One might be guided into this train of thought by remembering that the rainbow, which represents a gigantic spectrum effect, is considered to give similar indications of atmospheric conditions by the formation and preponderance of its colour bands; but we shall discuss this more fully shortly.

The Victorians were acutely aware of this effect, and much ingenuity was used in employing the spectroscope for the purpose of forecasting rain by studying the conditions of the upper atmosphere as portrayed by the instrument.

Upper air sounding was not possible in those days, and whereas the usual employment of hygrometer principles at ground level could not register vapour content at distance, such changes could be detected in the rain-band appearance in the solar spectrum.

The credit for identifying lines in the spectrum must be duly given to the great Munich optician, Fraunhofer, who produced a chart of the solar spectrum showing that it included hundreds of dark lines, some being more conspicuous than others.

To these lines he gave letters of the alphabet for ease of identification, and the letter D was dedicated to a prominent dark line in the yellow band. This line was later found to be due to the sodium properties in the light, so that we are able to say that if this intensity of the D line is found in any spectrum under observation, it is due to sodium vapour belonging to the body which is transmitting the light.

The indications mostly used to predict rain were the series of dark lines adjacent to the sodium D lines on their red border, increasing and becoming stronger with the increased water vapour content before the spectroscope.

Such use of the spectroscope seems to have been discarded, but it is likely that there is still scope for further research.

The rainbow

> 'I do set my bow in the cloud, and it shall be for a token of a covenant between me and the earth.'
>
> *Genesis 9, 13*

Just as the various colours which mix together to make white light are separated by the spectroscope, so a similar effect is achieved by nature in the production of the rainbow, which appears under favourable conditions when rain falls. The reason for this effect is that the raindrops are able to separate the sunlight into its component colours in much the same way as the prism does, and the total effect in the vastness of the atmosphere is a gigantic, beautifully coloured bow.

The complete circle is not seen from the ground but may be viewed from an aeroplane.

The principal colours observed are red, yellow, green, blue and violet, with mixed hues where the colours overlap.

Colour intensity is dependent on the size of the water drops present in the atmosphere. In fact, the drops can be so small that no colours at all are seen and the bow is whitish. The most brilliant colours are seen when the water drops are as large as they are during a heavy shower; smaller drops either reduce the intensity of some colours or fail to reproduce them entirely.

It is simple enough to reproduce a miniature rainbow for ourselves when we are watering the garden with a hose, since all we need are the conditions present during a good shower of rain. With our back to the Sun, we can throw out a fine spray in front of us, and by moving slowly from side to side we can find the position from which the rainbow (or waterbow) becomes visible.

When we have our back to the Sun during or just after the rain, we can see that the rainbow is centred exactly opposite, and often we can see two bows, the second being a little broader and a little fainter than the first. The original and brightest is called the primary bow, while the other is called the secondary bow and arises from a double reflection from the raindrops.

In the primary bow the red band is always on the outer edge, and in the secondary it appears on the inner edge, so that all the colours are reversed with respect to the primary.

The morning rainbow is always seen to occur in the west, usually with clear, bright skies to the east, and this indicates the advance of rain clouds from a westerly direction, as the appearance of the bow indicates a high percentage of moisture in the air.

When the Sun has passed to the east and the temperature begins to fall, a rainbow in the eastern sky with clear weather to the west tells that the rain has passed and that the next few hours, and possibly the next day, will be fine.

A rainbow appearing in good weather foretells bad weather to follow.

If only sections of the bow are visible, it may rain for several days.

When two colours are predominant, such as red and yellow, good, fair weather may settle in for a few days.

When, after a period of poor weather, blue is predominant, conditions will become more settled.

When the green band is seen to be brighter than the others, rain will continue, while if the green and blue bands are prominent with the other colours distinct, it may rain during the night.

If red alone is the brighter band, there will be rain and wind.

Yellow being predominant may indicate high winds.

Following long periods of wet weather, it is often the appearance of a rainbow that indicates fair conditions to come.

Rainbows are also produced by the light from the Moon

striking raindrops in the same manner as solar light, but owing to the comparatively feeble illumination from the Moon, the bow so formed is generally without colours, unless the night is exceptionally clear when the colours, in reduced intensity, are to be observed. This bow is called a lunar rainbow.

Although this chapter has been entirely devoted to the cause and effects of water vapour, and mostly the latter, we are now about to venture into the most exciting realms of its visual effects, namely the clouds and the part they play in our weather systems.

4/THE CAUSE AND EFFECTS OF OUR CLOUDS

> 'If a little cloud suddenly appear in a clear sky, especially if it comes from the West, or somewhere in the South, then there will be a storm brewing.'
>
> *Bacon*

Many similar statements dating far beyond Bacon's time prove to us that man has recognised the appearance of clouds as a means of predicting weather conditions, even in biblical times.

It is these cloud formations which we are now about to discuss, and it is of interest to note that the type and behaviour of clouds are positive indicators to the coming weather.

All clouds are composed of countless millions of water droplets which have condensed out from their previously invisible gaseous state when the temperature of the air in which they are suspended is cooled below its dew point, thus becoming visible as recognisable types of cloud formation.

Unless it is close to saturation point, air can absorb a considerable amount of water with no noticeable effect, and this absorption increases up to a maximum amount as the temperature is increased. The water vapour content is measured as grams of water per kilogram of air.

On a summer's day, air at 68°F can absorb a maximum of 15 grams of water per kilogram of air before saturation point is reached, but at 50°F, the air can absorb only half that amount.

At minus 50°F at 10,000 feet on the same summer's day, the air will absorb $\frac{1}{10}$ of the amount before reaching saturation point.

When the air is saturated and experiences the effects of cooling, then condensation takes place, and the height at which this happens and the type of weather system which is causing it will determine the type of cloud so formed.

During day mists in summer, we will all have noticed how, when the mist passes through a spider's web in the garden, thousands of tiny water droplets are collected on its fine threads, and adhere there, glittering like fairy candles in the sunlight.

This effect is simply the water content of the cloud which has formed at ground level when the temperature of the air has been caused to fall below its dew point by contact with the cold ground.

So you see, although we have progressed to the formation of our wonderful clouds, we are still, in principle, dealing with the effects of water vapour.

Identifying the clouds

The cloudy sky is an ever changing picture of shapes and colours, of clouds at varying heights and of clouds moving at different speeds; and frequently there are many different clouds to be seen at the same time. This dramatic cavalcade of nature provides us with a continuing story of the conditions of our weather.

Telling us this story are four basic cloud forms, namely

a. cirrus
b. cumulus
c. stratus
d. nimbus

Ten further formations are derived from these four basic groups. A combination of two of the basic forms sometimes represents a transitional stage from one type to another, and a prefix or suffix may be added, such as 'fracto' to indicate broken or ragged cloud, or 'lenticularis' which is applied to clouds having the shape of lenses.

It will also be noticed that the cloud formations are grouped according to the approximate height at which they occur, thus (Fig 5):

a. *High clouds* (cloud base above 20,000 feet)
   Cirrus (Ci)
   Cirrocumulus (CiCu)
   Cirrostratus (CiSt)
b. *Medium clouds* (cloud base between 8000 and 20,000 feet)
   Altocumulus (ACu)
   Altostratus (ASt)
   Nimbostratus (NbSt)

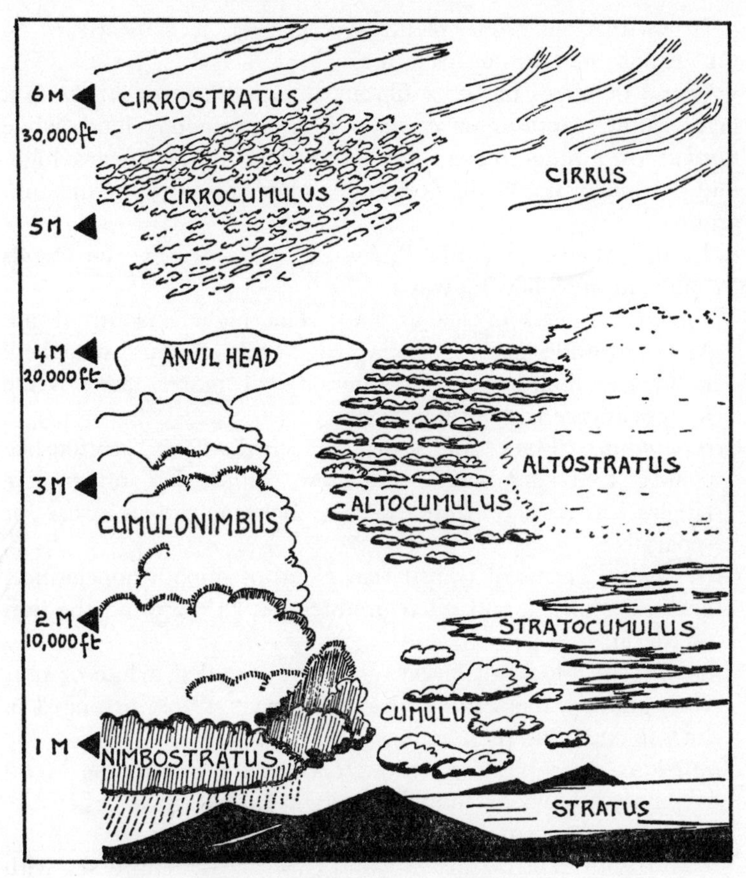

Fig. 5 *Cloud types and altitudes*

c. *Low clouds* (cloud base from almost ground level to below 8000 feet)
Stratocumulus (StCu)
Stratus (St)

d. *Convection cloud* (mean high 28,000 feet mean low 600 feet)
Cumulus (Cu)
Cumulonimbus (CuNb)

In the low cloud group we can get Fractostratus, and in the convection cloud group we can get Fractocumulus, indicating a breaking-up of these clouds. This term is only applied to cumulus and stratus.

In sorting out these picturesque names, it is necessary to understand the Latin terms in use.

Cirrus or *cirro-* means a fibrous cloud, *stratus* or *strato-* is a layer cloud, *cumulus* or *cumulo-* is a convection cloud, while *nimbus* or *nimbo-* means a raining cloud. *Alto-* means high, and is applied, not to the very highest group, but to the medium group.

In the order previously listed, we may identify the clouds by name in the following way:

Cirrus. Composed of ice crystals. The highest cloud of all. Mean altitude about 30,000 feet. Thin, fibrous, detached, moving at high speed. Sometimes called 'mares' tails'. White in appearance with a silky sheen.

Cirrocumulus. Flocks of small, detached, fleecy, cotton-like clouds, thin and without shadows, appearing rather like ripples in regular formations, either merged together or separate.

Cirrostratus. Forms of transparent cloud of smooth appearance, generally fibrous and often producing a halo around the Sun or Moon.

Altocumulus. Flocks of larger, generally rounded, white or partially shaded, masses of cumulus in layers. Often arranged in lines in one or more directions.

Altostratus. Grey-blue veil through which the Sun or the Moon is faintly visible as though through frosted glass. They are fibrous or uniform and may totally cover the sky.

Nimbostratus. A structureless cloud covering the entire sky with a grey layer from which either rain or snow falls, according to the season. The Sun is not seen through its depth and its colour may range from grey to dark grey.

Stratocumulus. Large globular masses or rolls of grey or whitish layer cloud with very dark patches, often covering the whole sky in winter.

Stratus. Overcast grey layer cloud of fairly uniform height which may give way to drizzle or snow. The Sun is sometimes visible through it. When broken and irregular it is called *fractostratus.*

Cumulus. Heap cloud. In great abundance during the warmest hours of the day. Thick and rounded, detached and dense with sharply defined outlines which can be seen to be developing vertically in domes and towers which look like large cauliflower heads. They are formed on rising currents of heated air and usually have a more-or-less flat base which marks the condensation level. The top portions are brilliantly white, being fully illuminated by the Sun.

Cumulonimbus. Massive thunder cloud, like mountains or turrets of great vertical extent, often having a screen of fibrous appearance above called false *cirrus,* and a mass of dense *nimbus* below, from which rain falls.

When the cloud is broken with rain, or parts of it become ragged it is called *fractocumulus.*

With the aid of a set of cloud pictures it is not difficult to recognise the different types of cloud, and for anyone about to take the matter seriously it would be advisable to write to the Royal Meteorological Society for a list of their publications on this subject.

In addition to the clouds listed above, there are a number of terms which assist us in identifying the *species* of cloud; that is to say that certain clouds have additional characteristics which are important enough to be noted by the inclusion of a descriptive term as a suffix.

Castellanus. In respect of cirrus, cirrocumulus, altocumulus and stratocumulus, the development of a series of vertical clouds arising from the main cloud in the form of castles or turrets of cumuliform development.

Calvus. In respect of cumulonimbus, from the upper regions of which almost vertical developments occur which are no longer quite of cumulus form and not yet of cirriform, but a whitish fog-like mass.

Capillatus. In respect of cumulonimbus in the upper regions of which it is possible to positively identify cirriform development. Often the form of an anvil can be distinguished and the formation may produce heavy showers of rain or hail, and thunderstorms.

Congestus. In respect of cumulus clouds which develop to great vertical extent with cauliflower-like protuberances in the higher regions.

Fractus. In respect of stratus and cumulus in which the cloud begins to break up at its edges and causes ragged break-away areas of cloud.

Floccus. In respect of cirrus, cirrocumulus and altocumulus, when the total cloud amount is present in small tufts of cumulo form with ragged undersides.

Fibratus. In respect of cirrus and cirrostratus in which the form is seen to be either in irregularly curved streaks or almost straight.

Humilis. In respect of cumulus clouds of horizontal rather than vertical extent.

Incus. In respect of a cumulonimbus which develops a smooth, fibrous anvil form.

Lenticularis. Principally in respect of cirrocumulus, altocumulus and stratocumulus, when the clouds form into an almond shape of fairly sharp definition. They arise mainly from the effect of orographic lift, which is the result of air being forced upwards by a hill or mountain.

Mamma. In respect of cirrus, cirrocumulus, altocumulus, altostratus, stratocumulus and cumulonimbus, in which the underside of the clouds develop heavy, bulging protuberances.

Mediocris. In respect of cumulus cloud of only moderate extent.

Nebulosus. Mainly in respect of cirrostratus and stratus in which the cloud development is like a nebulous veil.

Pileus. Mainly in respect of cumulus and cumulonimbus in which the rising air of a thermal (warm air current) reaches its condensation point and meets a layer of stable air which resists the attempts of the thermal to rise through it. Often the result is that the thermal forms its cloud in the area of saturated air just below the inversion (stable air layer), and then presses upwards to penetrate the air. This causes a cap-like dome to form, which is distinguishable from the main cloud and is known as a *Pileus.*

Spissatus. In respect of cirrus which appears greyish when seen through the sun.

Stratiformis. In respect of cirrocumulus, altocumulus and stratocumulus in which the cloud develops an extensive horizontal layer.

Uncinus. In respect of cirrus which forms hook-like tails or commas.

Virga. In respect of cirrocumulus, altocumulus, altostratus, nimbostratus, stratocumulus, cumulus and cumulonimbus in which signs of fallstreaks, or precipitation trails, are seen from the underside of the cloud, and which do not reach to the Earth.

Before it becomes possible to attach these identity tags to the basic cloud formations, it is necessary to make a close study of the subject using the sky as a practical laboratory.

Convection cloud

The process by which cumulus cloud is formed by strong rising currents of warm air is known as 'convection', and the current which causes the cloud is known as a thermal. Extensive developments of cumulus are caused by several thermals working in the same vertical direction so that the clouds grow upwards. Fine weather cumulus is relatively small, resembling little puffs of cloud floating in uniform lanes. Storm cumulus develops to enormous proportions until it becomes cumulonimbus.

When the Sun heats the Earth and the Earth begins to radiate its heat upwards into the atmosphere, it does so after the fashion of a large bubble which is full of turbulent air currents striking vigorously upwards, cooling, expanding to the outer edges of the bubble and falling to make room for more uprising air through the centre. All the time, the bubble is increasing in height until it reaches the condensation level of that particular air mass and begins to condense out into visible cloud, when the mechanics of its growth are shown in the convection of the cloud it is forming.

Other clouds are also caused by condensation, but instead of being directly due to rising thermals, they are due either to water vapour turning into ice, as with cirrus clouds, or to cold air undercutting warm air and forcing it up to condensation levels where it forms altostratus.

Orographic cloud

If the Earth had a smooth surface, the winds which circulate around it would move easily from one area to another being affected only by the rotation of the Earth. As it is, the Earth's surface undulates, drops sharply into valleys, rises steeply to make the hills and mountains, and generally provides an irregular obstacle-course for even the most gentle of winds.

As the wind strikes a steep hill, it is caused to change its direction from the horizontal to the vertical, which means that as it rises it cools rapidly, and in doing so may reach its dew point.

When this happens, puffs of cumulus form above the hill or ridge, and this is known as orographic cloud. Should this continue to build up and become heavy with water droplets, the rain which falls is called orographic rain. Altocumulus lenticularis is a form of orographic cloud.

The 'Levanter' is a good example of orographic cloud. It forms on the easterly winds around the Rock of Gibraltar, hiding the summit from view for days on end.

Anvil cloud

When rising cumulus encounters a stable layer of air, it is not always strong enough to penetrate it and is caused to spread out horizontally, rather like cigarette smoke meeting the ceiling and, being unable to rise further, taking a horizontal path at ceiling level.

In spreading out, the cloud often looks heavy with rain and forms the shape of a blacksmith's anvil. The cloud is correctly a stratocumulus which has been caused by the spreading of the main cumulus, but is generally called a 'water anvil', although rain does not necessarily fall from it.

The Nephoscope

A nephoscope is an instrument which measures the angular velocity of a cloud, and indicates its direction of movement.

The standard model as used at most synoptic stations is the Besson comb nephoscope. This instrument looks rather like a garden rake with widely-spaced teeth pointing vertically at the sky, and consists of a row of spikes 6 inches apart mounted along a horizontal rod which is fitted centrally to a vertical spindle. The spindle rests on a ballbearing base, so allowing it to rotate about a vertical axis under the control of the observer, who adjusts the position of the spiked rod by means of a pair of ropes attached to the spindle.

The direction which the observed clouds are following is indicated on a direction plate attached to the centre rod, while the speed of the clouds is found by checking with a stop-watch the movement of the clouds against the spacing of the spikes.

There are other types of cloud-measuring instruments but all are grouped under the generic term of 'nephoscope' and have a history dating to Goddard's 'cloud mirror' of 1851 and Besson's Comb of 1897.

Recording the cloud amount

We cannot fail to notice how the otherwise blue sky of day-time is sometimes quite obscured by low cloud, or that the Moon which we can reasonably expect to see at night is either hazily visible or not visible at all. We may note also that there are various degrees of sky or lunar visibility, according to how much cloud there is between us and the heavens.

To weather forecasters, the cloud amount present at the time of taking an observation is of importance since this information assists in completing the whole weather report, and gives an indication of the varying conditions within the weather system prevailing at the time. There is, therefore, a recognised method of observing and noting the condition of the sky, which, as we shall see in a later chapter, is transferred to the weather map by means of special symbols.

The 'OKTA' is the international unit for reporting the amount of cloud in the sky and, as a unit, is representative of $\frac{1}{8}$ of the visible sky area, the assessment of the total cloud amount lying between 0 for an absolutely cloudless sky, and 8 for an absolutely overcast sky. A final figure, 9, is used to explain that owing either to complete darkness, fog or smoke, an observation has not been possible. Each stage between 0 and 8 – that is to say, from 1 to 7 – represents a scale of values with which it is possible to use the sign + to indicate 'slightly more than . . . , and the sign — to indicate 'slightly less than . . .'

When extensive cloud formations show small areas of blue sky, it is permissible to use the term 'overcast with openings'.

The complete scale of sky cover together with plotting symbols is as follows:

Code	*Description*	*Symbol*
0	completely cloudless	
1	trace to 1/8	
2	1/8+to 5/16	
3	5/16 to 7/16	
4	7/16 to 9/16	
5	9/16+to 11/16	
6	11/16+to 7/8	
7	7/8 to 8/8 (overcast with openings)	
8	completely overcast	
9	impossible to estimate	
–	missing or doubtful data	

In all, the observations made of the clouds fall into five categories:

a. estimating the cloud amount;

b. recognising the type of cloud present;

c. estimating the height of the cloud base;

d. finding the direction from which the clouds are moving;

e. determinating the speed of the clouds' movement.

As a point of general interest, it should be noted that all cloud types may be found at greater heights in the tropics and lower heights in the polar regions. For example, cumulonimbus may

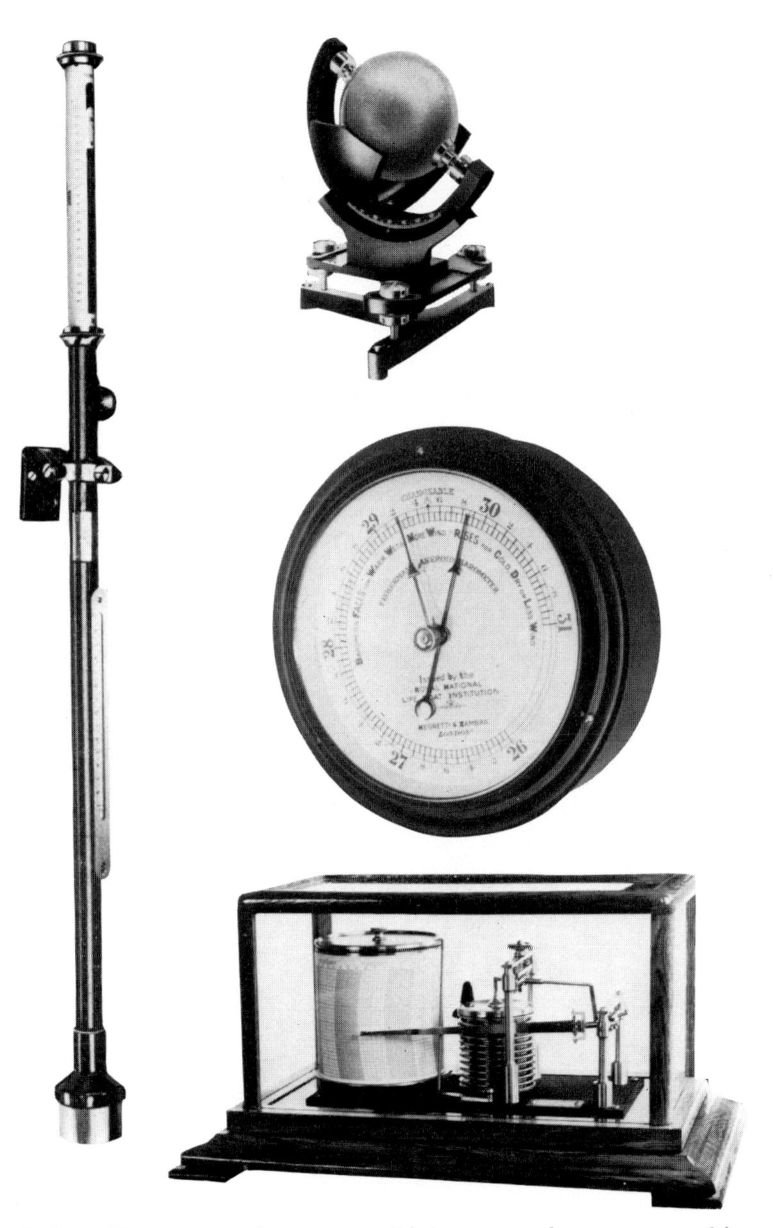

Left 1. Kew pattern barometer. *Right, top to bottom* 2. sunshine recorder; 3. aneroid barometer; 4. barograph

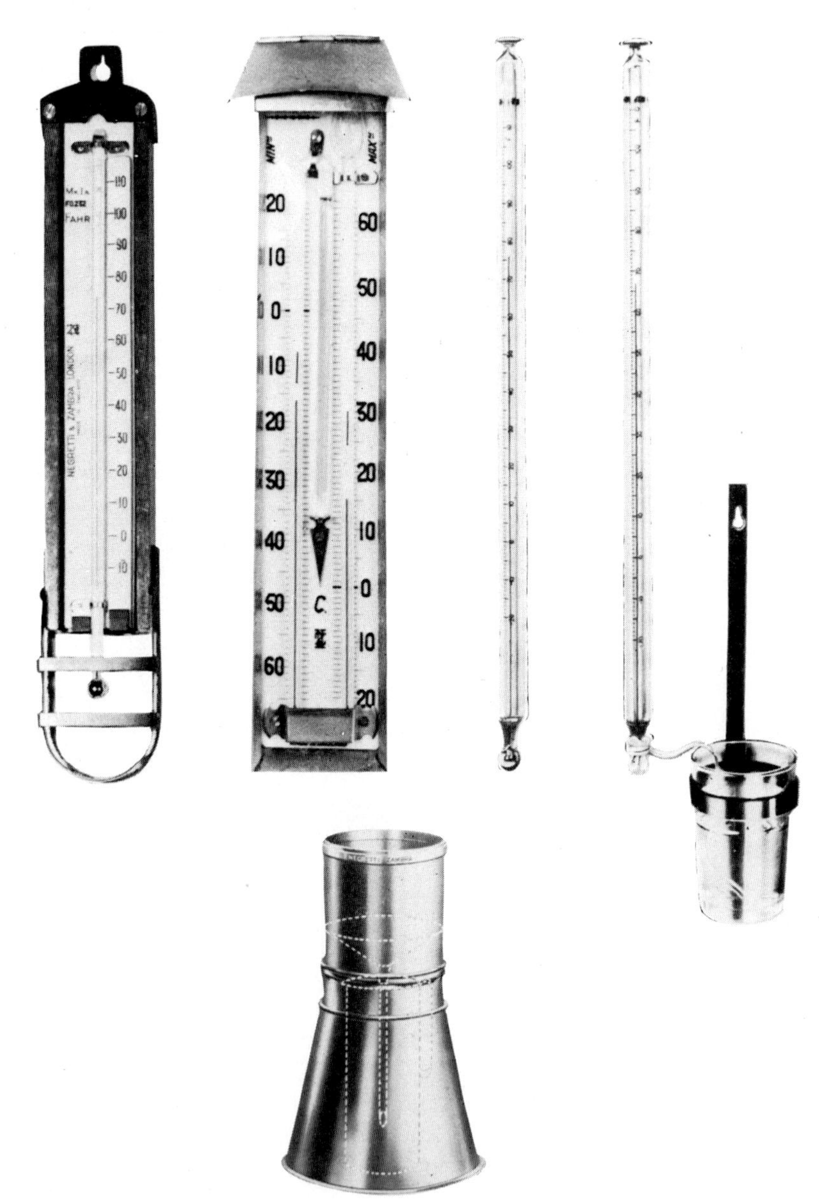

Above, left to right 5. air temperature thermometer; 6. maximum and minimum thermometer; 7. wet and dry bulb thermometer; *Below* 8. raingauge

9. Cirrus

10. Cirrus and cirrostratus ahead of bad weather

11. Shallow cumulus on a fine summer afternoon

12. Towering cumulus in an unstable atmosphere

13. Distant cumulus with anvil

14. Semi-transparent altocumulus exhibiting typical cellular structure

15. Altocumulus castellanus associated with thundery weather on a summer afternoon

16. Sun shining through a semi-transparent layer of altostratus

17. Stratocumulus on a summer evening

18. Mamma formation on the lower surface of a thick strato-cumulus

19. Wave cloud

20. Billows, or ripples, in thin altocumulus

21. The thunder cloud. Cumulonimbus

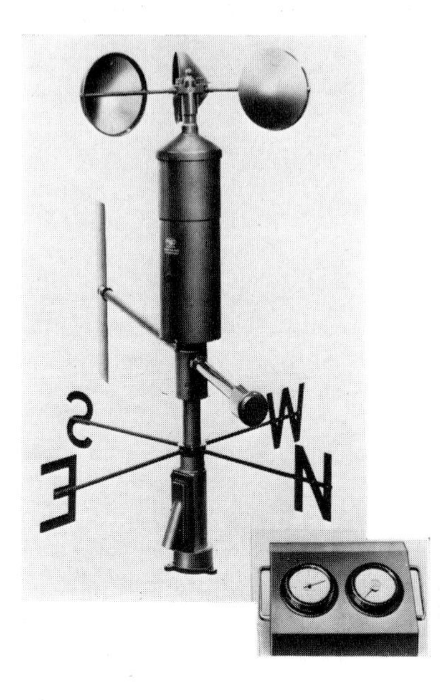

Left 22. Wind vane. *Right* 23. Anemometer combined with wind direction indicator. *Below* 24. A depression photographed from a weather satellite

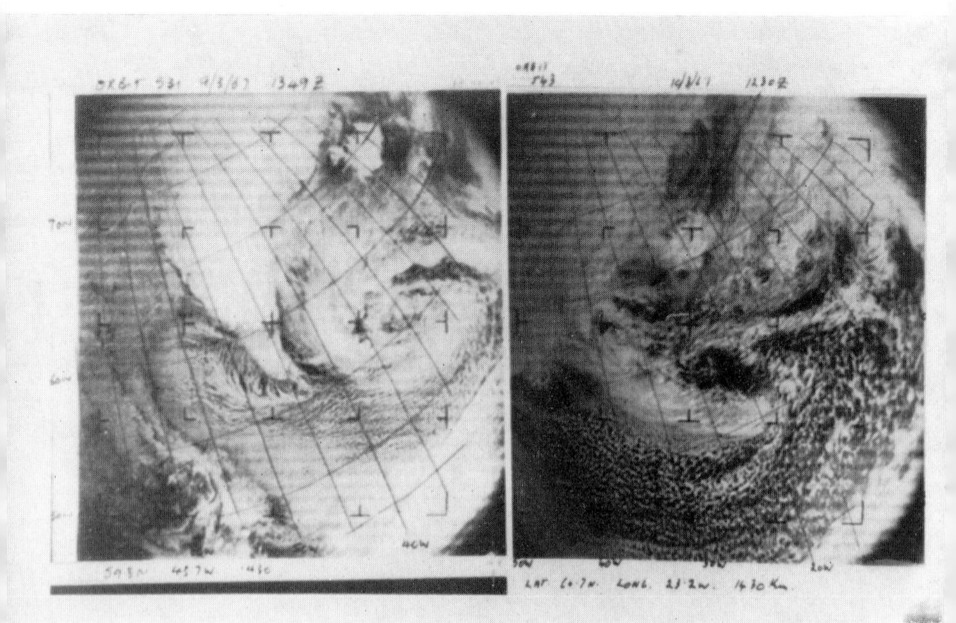

25. Earth from 22,300 miles in space

26. Artist's impression of a Geostationary Operational Environmental Satellite (GOES) in orbit

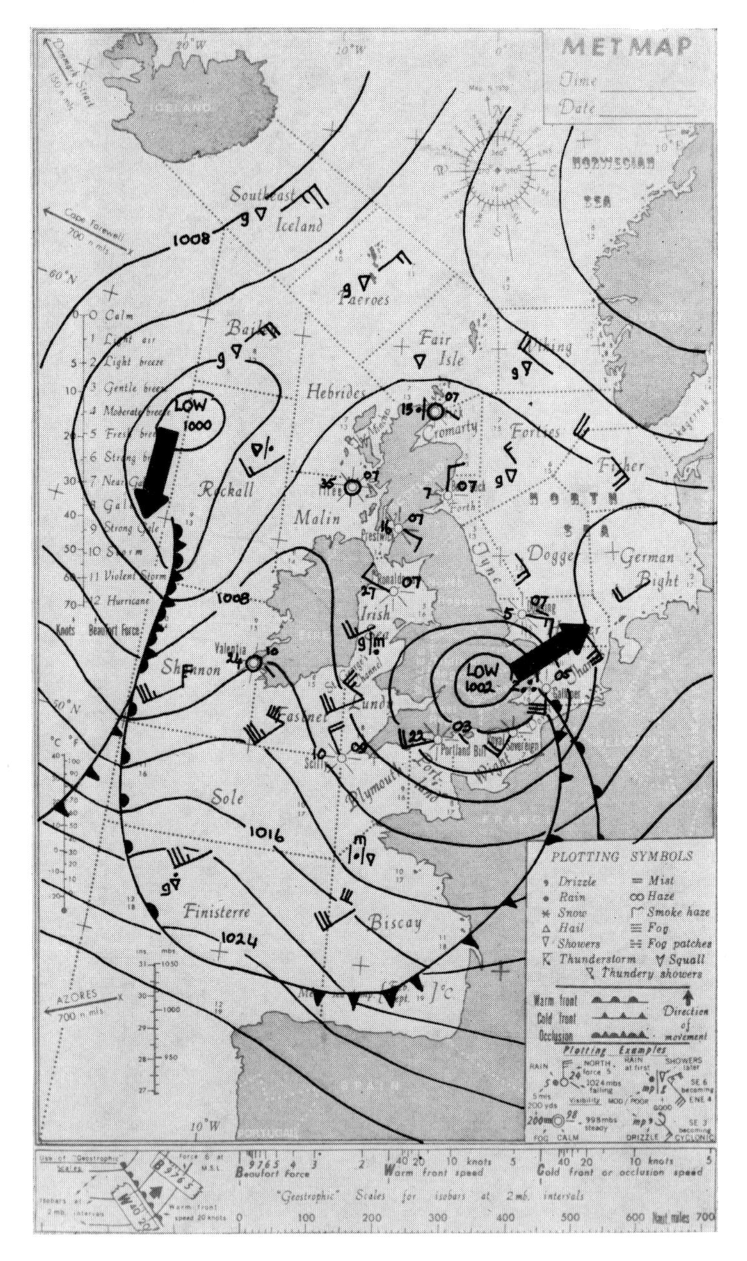

27. A plotted Metmap. Information from the Sea Area Forecasts and Coastal Reports have here been committed to a Metmap. After initial sketch lines have been made, this finalised version emerges from the forecaster's opinion of the situation

develop to 50,000 feet in the tropics, and cirrus may reach as low as 9800 feet in temperate latitudes, being even lower at the Poles.

What the clouds foretell

'Can anyone understand the spreading of the clouds?'
Job xxxvi, 29

The shape of each type of cloud is due to the circumstances of its origin and growth, and by studying the clouds it is possible to determine their origin and to forecast the coming changes of weather.

Cirrus

All cirrus clouds are composed of tiny ice crystals because at that height (20–30,000 feet), the air is well below freezing point. They appear, often moving rapidly from the west, as fine, hair-like streaks, commonly called 'mares' tails', or as clouds extending across the sky in long feathery streaks (Plate 9).

Sometimes, on a fine day when the upper air is calm, the cirrus can be seen in tangled masses. When cirrus strays across the sky alone and unaccompanied by other clouds, it is reasonably safe to forecast a dry, fine day.

The clouds should be well formed and moving in uniform motion and direction.

If the cirrus appears from a south-westerly direction and appears to be drifting, this may be taken as a sign of rain to come.

Should there be a fall in the barometer pressure at this time, one would be safe in expecting poor weather within a few hours.

Bands or streaks of cirrus occurring after a long spell of good weather may be the sign of a change. If these clouds approach from the windward quarter and develop into cirrostratus, this may be the forerunner of rain. A halo around the sun is produced by the refraction of the sun's rays through a layer of featureless ice crystals which is called cirrostratus.

Cirrus does not always cover the whole sky, but is sometimes seen rising from the horizon and finishing in a well defined edge. On the other hand, if the cloud dissolves into a clear blue sky, fine weather may be indicated.

In addition to the halo, there are sometimes other rings of coloured lights made visible by the effects of cirrostratus. One of them is a small bright patch of light located directly above the Sun, and a spot of light to one side of it, called a 'mock sun'.

Cirrostratus is often seen extending across the sky in long narrow, rapidly moving bands, which is due to the fact that at about 30,000 feet or higher, there is a band of winds travelling at something in excess of 100 knots. This is called the 'jet stream' and gives rise to parallel bands of clouds.

One of the first indications of an approaching depression is the appearance of cirrus with cirrostratus coming up close behind. It is the cirrostratus which produces the solar and lunar haloes, so that once one of these becomes visible, it is a sure sign of rain to follow, providing that the weather system keeps to its observed course (Plate 10).

Cirrocumulus

Otherwise called 'mackerel' sky, cirrocumulus is generally accepted as the forerunner of rain or wind, or both.

'Mackerel sky and mares' tails.
Make lofty ships carry low sails.'

If the weather on the day of appearance is fair to good, rain will be approximately 24 to 48 hours away.

When the edge of the clouds are frayed and roughened by the winds at their level, one may expect winds at ground level to follow.

When the threads are hooked and frayed, expect rain and wind.

The dense compressed clouds closely formed may herald thunder. Black, loosely-packed clouds drifting in summer may mean a storm.

Cirrocumulus and stratocumulus together may mean a storm.

In winter, the appearance of these clouds may indicate a change to warmer temperatures and possible rain.

'High upper clouds, crossing the Sun, Moon or Stars in a direction different from that of the lower clouds, or the wind then felt below, fortell a change of wind towards that direction.' *Fitzroy*

Such movement is also remarked upon by another observer thus:
'Clouds floating at different heights show different currents of air, and the upper one generally prevails. If this is NE, fine weather may be expected; if SW, rain.'

Cumulus

On a fine day, the cumulus clouds appear over the heated land masses and along the coast, separated from each other and not too large. This 'fair weather' cumulus tends to be of a uniform level with flattened bases and looks like cauliflower tops with horizontal rather than vertical development. The clouds tend to retain that form into the early afternoon. The term 'fair weather' is simply an indication that there is no evidence of rain cloud at the time of observation, and should not be accepted as a suggestion that rain cloud will not develop later (Plate 11).

Cumulus clouds are a pleasing sight as they drift in small puffs across the summer sky at medium height. Continued fair weather may be expected when these clouds prevail in the sky until quite late in the day before giving way to the evening stratus.

In summer, cumulus forms shortly after sunrise, dappling the sky with small woolly clouds. If the size and frequency of the clouds increases towards midday and clears by sundown, the calm conditions under which they formed (fine or settled) are likely to continue for another 24 hours or so, when the future weather will depend upon any fresh developments.

Should the cumulus develop towards evening and the sky become patched with other cloud forms, then rain may come within 24 hours.

If the clouds evaporate, they take on a dull colour and become ragged as they break away, while the rising tops have white, sharply defined outlines.

A developing cloud progresses from small puffs into a larger formation with a flat base, a phenomenon due to the fact that the rising warm air causing it all cools at the same rate and reaches condensation level at the same time.

Development of cumulus into towering, rolling masses with brightly outlined edges will cause precipitation of snow, hail or rain, depending upon the season and temperature (Plate 12)

After showery weather, it may be seen that the tops of the cumulus will not grow any higher and that its upper regions are spreading horizontally, in which case it is unlikely that it will develop into a shower cloud.

When, in summer, there is a sky full of clouds at all levels, and cumulus, altocumulus and cirrus are all visible at the same time, and all travelling away at different speeds and in different directions, it is a sure sign of thundery weather.

Cumulus over the sea is more noticeable in autumn and winter because the sea area is often warm in relation to the land area, making for a good deal of convection and rising thermals.

During the summer there is little cumulus formation over the sea due to the fact that the water is absorbing the Sun's heat to a considerable depth, and is not radiating it back into the atmosphere with any noticeable effect. At night, when the land rapidly loses its heat, the sea, being the warmer of the two, radiates its heat into the air and may form cloud.

Cumulonimbus

Whichever way one looks at this cloud formation, it means thunderstorms. Did Shakespeare refer to this in *The Tempest* when he wrote:

'And another storm is brewing; I hear it sing i' the wind. Yond' same black cloud, yond' huge one, looks like a foul bumbard that would shed his liquor . . . Yond' same cloud cannot chuse but fall by the pailfulls.'

With some justification, the cumulonimbus is otherwise referred to as 'The Thunder Cloud', differing from the ordinary cumulus in as much as that it is considerably larger and extends to altitudes through several miles. Its rolling, turbulent domes rise majestically and cast heavy shadows. It has a low base and towers to the highest points of all.

The cumulonimbus can develop from comparatively small clouds which are caused to grow and become more turbulent by the rising air currents causing them. From this, the large cumulus can develop, and begins to fill the sky. Eventually, deep turrets or castles form, indicating that showers are not far off.

As the cloud increases in height, the top may lose its heaped appearance as it develops the nimbus which will manufacture the forthcoming rain.

Observing the top of such a cloud, one may see how the heaped formation slowly reforms into the shape of an anvil head, composed of thread-like cloud. At this stage our one-time harmless little cumulus has now become a threatening cumulonimbus from whose lower regions heavy rain will fall (Plate 13).

Sometimes a thin band or funnel of cloud can be seen depending from the base of the thunder cloud. These funnels are like small tornadoes as they follow on behind the parent cloud, twisting and turning like demons, leaving a trail of wreckage in their wake.

Nimbostratus

'Neptune brings his blanket overhead,
To wring it dry to put abed.'

Frequently seen from about October and throughout the winter, the nimbostratus brings the dull, gloomy, overcast type of atmosphere which depresses everyone and will at intervals spread rain over large areas. The only sign of relief from this rain is when the cloud becomes lighter as the Sun penetrates its mass and makes it thinner by evaporation.

The honeycomb or webbed type of stratus will also bring rain with dull skies, and when we observe high, wide-spread clouds drifting in the opposite direction above low-level, darker nimbus, we may well forecast a spell of bad weather.

During winter months, the nimbostratus can bring continuous snow.

There are also other signs which are worthy of note in conjunction with the appearance of cumulonimbus. For example, if one observes numerous high clouds spreading out in different directions well above the approaching cumulonimbus it is likely that the coming rain will be long-lasting and heavy.

Conversely, it is reasonable to suppose that if the upper clouds are few, then the rain will be light and short-lasting.

Following close behind the nimbus, and accompanied by a high wind, will be seen a ragged, ungainly cloud which is usually called a 'rag cloud'. If it diminishes and breaks up, then the weather is clearing, but if it develops and becomes thicker, then is the time to look out for gales to follow.

At sea, the approach of streaks of nimbus from the south-west is taken as a sign of wind.

Often, where there is a high wind accompanying the cumulonimbus the whole cloud is torn apart and breaks away in shreds, when it is called *fractonimbus.*

Altocumulus

The clouds of altocumulus are made up of a great many globules of cumulus-type cloud, sometimes to be observed at two distinctly separated levels, the upper level being at approximately 16,000 feet, and the lower at approximately 10,000 feet.

The type of altocumulus which is like cotton-wool puffs, or flock, is called *altocumulus floccus,* and is sometimes seen to precipitate showers of minute snow crystals, while its speed of passage across the sky will indicate an increase of wind speed with increase of height.

The *altocumulus lenticularis* is a large, cigar-shaped cloud often seen in the evening, and is so named because of its similarity to a glass lens. Although these clouds may appear over areas without any apparent up-currents, they more frequently appear as the result of marked orographic origin. Rain may therefore be expected to fall as the result of cooling.

Like all other clouds, and depending upon conditions, the altocumulus can either disperse or develop or change shape into other cloud forms.

In developing, it may form with its masses of cloud globules arranged along a relatively flat base, while the upper regions take on a turret-like formation. It is then called *altocumulus castellanus,* and may warn of thundery, wet weather (Plate 15).

In changing, the cloud may join up with layer cloud and become *altostratus.*

Altostratus

When this cloud forms in such a way as to impart a watery appearance to the Sun or Moon, it usually indicates the approach of rain, and as soon as rain begins to fall, the cloud is re-categorised as *nimbostratus* (Plate 16).

Stratocumulus

These are layer clouds, (Plate 17) usually so extensive as to cover most, if not all, of the sky. They are layer clouds in the form of globular masses with heavy temperature inversion immediately above them, so that vertical development does not take place. When this cloud is seen in abundance in the north-east and south-east with a prevailing east wind, one might expect rain or snow, depending upon season and temperature. The white protuberances which can be formed on the underside of this cloud is called *mamma* or *mammatocumulus,* and indicates wind, continual rain, stormy and thundery weather (Plate 18).

Stratus

> 'If two strata clouds appear in hot weather to move in different directions, they indicate thunder. If during dry weather, two layers of cloud appear moving in opposite directions rain will follow.'
>
> *C. L. Prince*

Stratus is a layer of featureless grey cloud, usually of fairly uniform density at the base and increasing upwards.

It is caused by the vapours which rise during the day and which fall earthwards with the falling temperatures of evening, and because at this time the cooling of the air begins at ground level and slowly rises, the first appearance of the stratus is close to the ground from whence it increases upwards as successive layers of the lower atmosphere are caused to reach their dew point.

As this cloud generally appears around sunset and grows widespread during the night, dispersing at sunrise, it is often called the 'night cloud'.

The stratus is therefore associated with the mists which form during calm evenings at the bottom of valleys and across low-lying ground.

When morning comes and the Sun's heat is directed at the upper regions of the stratus, it begins to develop some turbulence and form into small cumulus. At the same time, the heat being radiated from the ground presses the base of the cloud upwards, so that it all ascends, and finally breaks up and disperses.

The presence of morning stratus is usually a sign of fine conditions to come, especially if the stratus dispels rapidly in the spring and summer months. It is also an indication of the continuance of existing fine weather.

However, stratus is more readily recognised as a sign of bad weather in autumn and winter, and appears as a grey cloud, giving prolonged drizzle and sometimes fog.

Wave clouds

Where the direction of the air takes it across a hill or range of hills, it is forced to rise in following the contours of the ground and as a result is raised above its condensation level, whereupon it forms clouds (Plate 19).

These are called wave clouds because, in passing over the hills, the air produces successive developments, or crests, of cloud which are separate and which take on a wave-like appearance with a smooth outline.

The clouds are more numerous at sunrise but, owing to their shallow formation, quickly dissolve as the Sun evaporates the water droplets of which they are composed. This is particularly so with inland formations, while at coastal areas the wave clouds tend to maintain their development for longer periods.

Wave clouds which have appeared with the rising Sun and have dissolved before the increasing temperatures of the day, may return again in the evening.

Billows

These are due to the action of convection within a thin layer cloud which breaks the layer in to cloudlets, forming an appearance like altocumulus arranged in wave-like patterns. Such cloud which breaks the layer into cloudlets, forming an appear-same speed as, and in the same direction as, the wind at their level, so that there is a continuing process of fresh billows forming at the near edges and evaporating at the lee edges of the general formation (Plate 20).

Before passing on to other sections of our book it should be mentioned that the study of clouds is as extensive as the study of any other facet of meteorology, and that many experts have devoted whole volumes to the theory of cloud development and to their use in practical meteorology. What we have discussed in this chapter is really little more than a non-technical summary of the whole subject.

Storms

> 'He causeth the vapours to ascend from the ends of the earth,
> He maketh lightning for the rain;
> He bringeth the wind out of His treasuries.'
>
> *Psalm cxxxv, 7*

> 'Behold, there ariseth a little cloud out of the sea, like a man's hand . . . Prepare thy chariot, and get thee down, that the rain stop thee not. And it came to pass that the heaven was black with clouds and wind, and there was a great rain.'
>
> *Kings xviii 44, 45*

Without the 'thunder cloud' there would be no storms, and so our familiar friend, the cumulonimbus, continues to occupy our attentions as we delve into the origins and effects of the weather condition that we generally refer to as a storm.

Thunderstorms usually follow a period of hot, dry weather, when the air at ground level becomes damp and cools rapidly with increase of height. The reason for this is that when an ordinary cumulus cloud is forming in stable air, the temperature decreases at the lapse rate of about 5°F every 1000 feet, until the point is reached when saturation takes place and the water vapour eventually turns into liquid drops.

If, during the ascent, the cooling drops below 3°F, the damp air will be caused to rise freely and thunderstorms will develop.

At this point we must recall that the thunder cloud differs from the ordinary cumulus in as much as it is considerably larger and extends to altitudes through several miles. Its rolling, turbulent domes rise high and cast heavy shadows. At the top of the cloud there will develop a broad sheet-like area which is easily recognised as the anvil shape it assumes (Plate 21).

The formation of the cumulonimbus indicates that thunder storms or rain will not be far away.

The cause of this upheaval in the atmosphere is a battle between air of different temperatures, since, when large areas of different temperatures meet, they may give rise to thunderstorms.

This meeting often happens at cold fronts. In this case, cold air is travelling at great velocity at higher altitudes than it is at ground level, and forces itself in above an area of warm air; the consequence of this is a sudden increase of the lapse rate and a disturbance of convection which results in thunderstorms.

When air at ground level becomes heated, it may rise rapidly and by so doing upset the general equilibrium of the air above it, resulting in heavy rain. This type of storm usually occurs durthe afternoon or evening.

If the barometer reading is fairly high and the air is dry, it is unlikely that any storm will arise, and there will only be cumulus formations.

If the barometer is lower than average, then a thunderstorm is likely, especially if the temperature falls as well.

It is the lapse rate of the temperature that decides the severity of the storm. Conditions are said to be stable when the normal fall of temperature is about 5°F in every 1000 feet, but the air is rendered unstable when the fall is more than 3°F in every 1000 feet of ascent, for the unbalanced conditions leave an abnormal layer of cold air which has a tendency to sink on a layer of warm air which is trying to rise.

Should this happen on a hot day, the warm air will ascend rapidly taking with it considerable quantities of moisture, and when this reaches condensation level, rain will fall.

The warmer the general conditions, the greater the capacity of the air to contain the moisture, and it is therefore quite reasonable that we should expect our most severe thunderstorms during the summer.

Usually thunderstorms are preceded by heavy, muggy air, and the barometer reads low against a comparatively high thermometer and increased humidity. When the rain comes, the temperature is usually caused to fall and one experiences a pleasant freshness of the atmosphere.

When the falling rain raises the humidity, the uncomfortable, moist warmth returns, but the approach of the cold front will clear the air and bring a fresh breeze.

More often than not, directly before a heavy storm, the air becomes very still and heavy and the birds stop singing – a prelude to the impending violence – the quiet before the storm.

Thunderstorms are not necessarily of a widespread nature; for example, air may be forced up a hill or mountain, cooling rapidly in its ascent, and the result can be in the nature of a heavy local thunderstorm.

Lightning

The movement of the water droplets within the immense cumulonimbus is violent and erratic. The drops are whirled around from lower to higher regions of the cloud, broken up, hurled through rapidly varying temperatures, they collect more water, become larger and form themselves around dust particles which become their core as they ascend and descend in continuous motion for the entire growth and life of the cloud of which they are a part.

This movement within the cloud generates negative and positive charges of electricity, in much the same way as charges are generated when a glass rod is rubbed with a piece of silk.

Because the charges are not travelling through a continuous circuit such as occurs when the household electricity supply is switched on for lighting or cooking, they are called static charges, and remain in suspension as either a group of positive or a group of negative charges for as long as conditions within the cloud will support them in their separate static states without allowing them to unite.

By way of clarification, we must mention that two *positively* charged bodies will repel each other, and two *negatively* charged bodies will repel each other, but a positively charged body will *attract* a negatively charged body and the two will attempt to unite.

Therefore, when there are positive and negative charges about, they will continually attempt to meet and to mix. The factor which determines precisely when their meeting will take place is the strength of the insulation between them, and the ability of the charges to increase their strength enough to overcome the insulation.

In the case of clouds, the air between them, or the air between the static charges which have developed within them, is the insulation. The upper regions of the storm cloud contain negative static electricity, and the lower portions contain positively charged particles; as yet unable to meet because of the insulating air gap between them.

The actual charge is brought about by the fact that as the raindrops are split up, those descending receive a positive charge and those ascending receive a negative one. The general tendency is for the more positive ones to assemble along the base of the cloud and for the more negative ones to continue to its upper extremities, leaving lesser, stray positive *and* negative charges distributed throughout the remainder of the cloud trying to find a place for themselves.

When the separately assembled charges have accumulated and reached great strength, there is a difference of potential within the cloud. Then the two great charges, overcoming the resistance of the insulation between them, join up and an enormous spark, called lightning, jumps across the cloud as the charges unite.

Lightning is not necessarily confined to only one cloud, since the discharge may occur between the difference of potential in two clouds, the positive charge of one joining up with the negative charge of the other.

On the other hand, when the Earth is the most conveniently charged body, the discharge will take place between cloud and Earth.

Whichever the arrangement, the result is the same – lightning discharging at millions of volts.

When the flash is seen direct, it is called 'fork lightning', and is resulting from nearby storms. When only the reflection is seen among the clouds, it is called 'sheet lightning', and the storm is some way off.

That lightning which discharges down to Earth is the most dangerous, for it may cause widespread damage and loss of life.

Lightning will strike at the highest point above the ground in order to find its way to the Earth. This is why the practice of sheltering under a tree is dangerous, and for the same reason one should not remain in any open space where one might present an outstanding target during a storm. It would be wiser to shelter in a forest of trees rather than use one single tree or small group of trees in an unsheltered spot, since the single tree is more conspicuous for lightning than is a whole forest of similarly sized trees.

A lightning conductor is used on many buildings and chimney stacks to conduct the electric discharge to Earth should it strike in that particular place.

All observed lightning is not accompanied by thunder as is the case with the fork and sheet variety. There is also the silent-lightning or heat-lightning, which occurs mostly during the serenity of a summer evening and illuminates the heavens for considerable periods. In the more northern parts of the British Isles, this type of lightning is considered as being the forerunner of storms.

The reason for the silence of the discharge must be that it is the reflection of lightning discharges from far distant storms in the upper regions of the atmosphere and that we do not hear the thunder over the distance separating us from the source.

The much-used saying that lightning never strikes twice in the same place is quite untrue, since the great spark may strike any one spot any number of times, providing there is sufficient reason for it to do so and that its course is directed over that spot.

It is probable that the saying arose through the fact that whatever the lightning struck would be destroyed and therefore would not be sufficiently prominent to be struck again, but, to take just one popular example, the beautiful church at Week St Mary in Cornwall, has been damaged no less than six times since 1688.

Not all old sayings are inaccurate, and the old Wiltshire jingle

'A storm to make the cows low
Will surely make the wheat grow.'

is one which has the ring of truth about it because it foretells that crops derive certain benefits from storms.

The explanation of this must be found in the amount of ammonia and nitrogen oxide which is manufactured by the lightning spark as it burns its way through the air. One may often experience the scorched acrid aroma during a storm after lightning has been close overhead, and this is the sign that conditions are right for the forming of this aerial fertilizer which nourishes the vegetation.

Mixing with the rain, the ammonia becomes a chemical

fertilizer, known as ammonium hydroxide, which breaks down with the nitrogen oxide and forms into nitric and nitrous acids on natural mixing with the water vapour and the air.

The detection of storms

As lightning flashes are discharging electricity, they emit waves at radio frequency, and these are often detected by the domestic radio receiver in front of an approaching storm and very clearly as sharp 'crackles' when the storm is overhead. By using special Radio Direction Finding equipment tuned to these discharge frequencies, it is possible to locate storms up to about 2000 miles distance, and to plot their path on a prepared chart.

Telephone communication between observation stations and the Control Station provides information concerning the direction of the lightning flash in relation to the observers, and this is plotted on to a chart at the control station, in much the same way as we are able to use direction-finding apparatus to pin-point the position of an aeroplane by listening to its radio signals and tuning to them on one or more receiver stations.

St Elmo's fire

> 'Last night I saw St. Elmo's stars,
> With their glimmering lanterns, all at play
> On the tops of the masts and the tips of the spars,
> And I knew we should have foul weather today.'
>
> *Longfellow* (*The Golden Legend*)

During storms it is probable that one may observe a faint glow around the top of lightning conductors, or around the masts on ships, or along the wings and fuselage of an aeroplane in flight. And there may be a steady charge of some length and duration, during which time the fire is dancing and flickering about the conductor to which it is attached.

This phenomenon takes place when a well-charged cloud passes close to the Earth, or aeroplane or ship, and the charge combines with that which is already being sustained by the object, only it combines in a more gentle manner than during a normal lightning discharge. It does not spark or rent the skies, but appears to slowly burn like a ghostly flame until it disappears.

The Aurora Borealis

While we are on the subject of electrical discharges we are bound not to forget the magnificent display of coloured lights which occur in the northern hemisphere, when they are called the Aurora Borealis, and in the southern hemisphere, when they are called the Aurora Australis.

Aurorae are usually green or blue-green with occasional patches or fringes of pink and red, and they vary in formation and intensity.

The fluorescent luminosity of the Aurorae is produced by the interaction of atoms and molecules in the upper atmosphere which are the result of electrical charges projected into space in great quantities by the Sun.

Some of these charges, which appear more especially from those areas of great sun-spot activity, travel across space in other directions, probably presenting aurora appearances to other planets, while some proportion of the charge crosses the path of the Earth.

When the Earth's position in space is reached, the magnetism propagated by the Poles attracts the charges from the atmosphere and concentrates them around the North and South Poles at heights varying between 55 and 80 miles above the surface, and occasionally 300 to 600 miles up.

The disturbances of the Earth's magnetism caused by this suspended concentration of electrical charge, excite the separating air and illuminate it for some hours with shimmering rays, moving and mingling in the heavens.

The appearance of a strongly-coloured Aurora may be accompanied by a magnetic storm, but this is not inevitable.

The effect, which may be expected at the time of the equinoxes, and when sun-spots are in abundance, begins with a low arc of light across the horizon and develops into curtain-like rays of golden yellow, brilliant white patches and streaks of green and crimson. At other times, the appearance may be quite unspectacular – nothing more than a shadowy glow – or may resemble many searchlights streaking upwards in a shimmering, changing display of colours.

Thunder

As the powerful spark of lightning discharges with a force in the order of 100 million volts, it causes a violent disturbance in the atmosphere by suddenly raising the temperature of the air, which immediately expands with incredible speed. When the flash has passed, the air at once cools to its previous temperature and contracts, rushing in to fill the vacuum left by the lightning. The result of this contraction is thunder; sometimes sharp and staccato, and sometimes rolling and echoing among the clouds like an enraged giant venting his wrath upon the skies.

The lightning spark may be fairly short, which produces the short thunder; or it may be some 2 or 3 miles long, which gives rise to an extended peal owing to the fact that the thunder is produced at all points along the line of discharge at practically the same time, but the sound, travelling at only about 1100 feet per second (1 mile in 5 seconds), takes time to reach our ears.

By noting the variation of this time, it is therefore possible to calculate the approximate distance of the storm by counting the seconds and allowing one mile of distance for every five seconds.

By calculating the interval between the flash and the thunder in this way it is possible to note whether the storm is approaching, receding or crossing by sound-detecting the direction from which the thunder comes, and by observing the change of position of the flash.

'Thunder in the morning denotes winds, at noon, showers and in the evening a storm.'

The Storm Glass

The Storm Glass or Camphor Glass, as it is sometimes called, is mentioned here more out of interest than for serious meteorological purposes, as there has been much controversy concerning its use in the past. Although there appears to be a large number of such glasses in use by amateur weather-watchers, it is not a professional aid to forecasting, but it is a fact that ancient mariners would stand by the indications of their storm glasses with greater conviction than many a modern meteorologist would support the readings of his barometer.

Nevertheless, it is suggested that the student may derive some pleasure and interest by experimenting with this simple indicator and, further to this, he may also be impressed by its accuracy in recording local conditions.

The glass is correctly a hermetically-sealed phial about 6 to 12 inches long. Its contents are varied, but the lists shown below will give cause for experiment and observation until a workable solution is found.

The formulae should be accurately mixed so that the ingredients are well proportioned.

Formula 1

Part A: 10 dr Camphor and 44 dr Alcohol.

Part B: 2 dr Nitrate of potash. 2 dr Ammonium Chloride. 36 dr Distilled water.

The mixture

11 dr of Part A with 7 dr of Part B, or 18 dr of Part A with 11 dr of Part B.

Formula 2

2 oz Distilled water. 2 oz Alcohol. 2 dr Camphor. 1 dr Potassium Nitrate. $\frac{1}{2}$ dr Ammonium Chloride.

Formula 3

Camphor gum dissolved in Alcohol with Nitrate and Sal Ammoniac to the satisfaction of the experimenter, based on experiments.

The mixture should then be poured into the glass phial, leaving two or three inches of the sealed end filled with air.

The sealing does not seem to be over-important, so that if it is not possible to complete the glass-sealing at the top, a cork fitted firmly into the open end will prevent foreign bodies spoiling the solution.

Generally speaking, the instructions are thus:

A In fine weather the crystals are said to settle at the bottom of the tube, leaving the upper part of the liquid clear.

B In stormy weather the crystals are said to rise, making the liquid cloudy and turbid.

C The approach of wind can be foretold by the clouding over of the solution after a fine spell. Heavy clouding will precede winds at gale force.

Some say that this action is due to variations of the mixture with wind direction when the glass is exposed to the air. Others say that it is due to the intensity of the light or variations in temperature.

In terminating this chapter, it is well worth recalling that we started out in Chapter III to discuss the cause and effects of water vapour, and that we have traced the life of these minute drops of water in most of its various forms; as an invisible gas held in suspension, as visible steam, as condensation, as dew, frost, snow and ice. We have seen it as rain and fog, as aerial colours in the rainbow, and, of course, as cloud, and, lastly, we have studied the part it plays in the production of storms. All this from little drops of water.

It is the nature of most of us to accept the variations in our weather as they occur; it is important, however, that the weather-watcher should be aware of the fact that his weather does not just happen in an isolated spot above his head, but that it has travelled great distances to get there and is the result of the movement of enormous air masses from various parts of the Earth's atmosphere, and that they, in turn, are due to vast differences of atmospheric pressure and temperature.

In order to gain a workable understanding of these massive world-wide movements, we must now turn our attention to the origination of the various air masses and to their circulation around our aerial ocean.

5/OUR RESTLESS AIR

'Ever moving here and there,
Whither way, O restless air?'

An understanding of the weather would be much simpler if it were to conform to a uniform pattern of behaviour all over the globe, but there is no yard-stick by which to determine the reason for an area of calm weather to suddenly develop a raging storm, or, for that matter, to predict the direction it will take and the speed at which it will travel; nor are we able to tell exactly if or when it will change its direction and advance on some other course, for, without apparent reason, our weather changes its character and its form, its direction, and speed, and its often unobliging temperament without giving us much warning of its intentions.

The climatic pattern of large areas of the globe do enable us to chart positions on the world map which are predominantly dry or predominantly rainy, but even these change their centres according to season.

In the tropics there is no seasonal sequence but a fairly regular pattern of weather affected mainly by the difference between winter and summer conditions, which produce either expected periods of rain or expected periods of dry weather.

Based on this knowledge, it is possible to show the expected state of these climatic areas during the periods January to June and July to December, in which we see mainly sunny areas and mainly rainy areas, although it does not guarantee that either state is continuously stable.

In temperate latitudes, on the other hand, the weather is changeable and difficult to predict; so much so that the weather of the British Isles continues to confound even the most experienced forecasters.

There is, however, the pattern of the four seasons with their typical weather systems and their rise and fall of temperature, their periods of anticyclone activity and their average behaviour towards stormy days and calm days.

This average behaviour is modified, if not governed, by the great air masses which are continuously travelling through the atmosphere and whose composition with regard to humidity and temperature are reasonably constant.

An air mass may cover an area of a million square miles or so, and its effect on the weather at a particular point on the globe is determined by the physical characteristics it has acquired at its source and by the modification it has absorbed during its journey.

Air masses

There are two basic air masses, that which is called Tropical and that which is called Polar (Figs 6a and 6b).

The general character of the Tropical group is warm and moist, bringing fine, clear weather in summer, and mist, fog and drizzle in winter.

The general character of the Polar group is cold and dry, with convection, causing squalls and gustiness. Cumulus cloud may form, with weather fair to showery and the possibility of thunder.

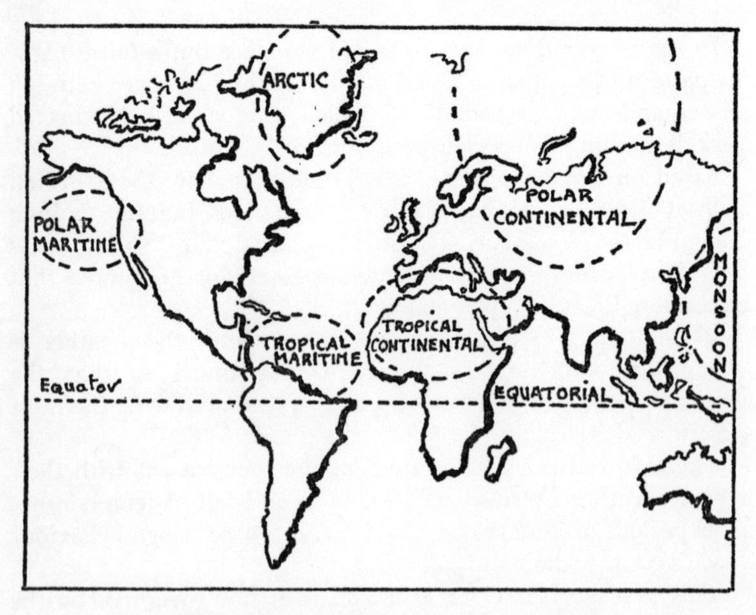

Fig 6a *World distribution of air masses*

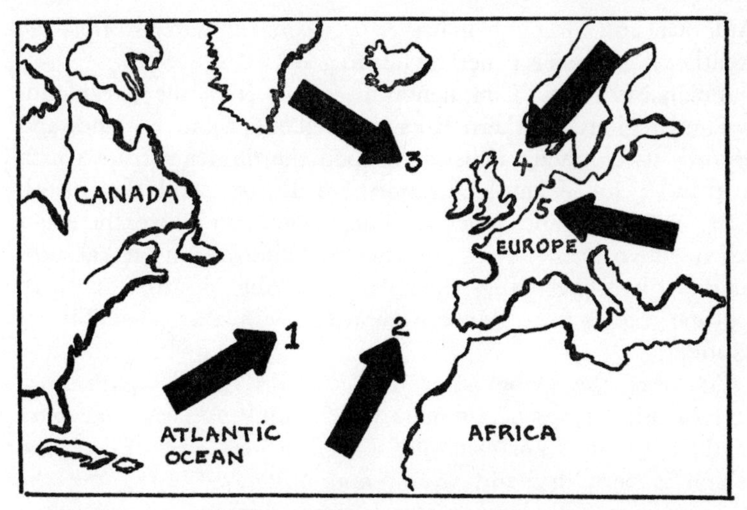

Fig. 6b *Where the weather is born and its effect on the British Isles.* 1 : carries frequent storms across the Atlantic. 2 : brings fine weather from the south. 3 : brings biting winds in winter. 4 : brings warm weather in summer and intense cold in winter. 5 : brings seasonable winter weather and cool settled weather in summer.

The two air masses each divide into two main types which give them their characteristics, namely Maritime and Continental, so that one may refer to Tropical Maritime or Tropical Continental air, and to Polar Maritime and Polar Continental.

Tropical Maritime air comes in from the sub-tropical Atlantic in winter, often bringing an unpleasant muggy atmosphere and prolonged drizzle. In summer, it brings us, from the Mediterranean, the fine weather which originated from the well-known anticyclone over the Azores, and carries high temperatures and brilliant sunshine. The fact that the air mass also brings higher humidity than other air masses may give rise to sea fogs in the Channel and along the coast.

Tropical Continental air is drier than Maritime air, and comes mainly from North Africa and south-west Europe, frequently producing heavy storms along the Mediterranean.

Polar Maritime air originates over the west coast of North

America, and although it brings occasional thunder storms, the weather it gives us is generally fair to good.

Polar and Arctic Continental air comes from the Greenland-Iceland regions, northern Russia, Spitzbergen and Finland, and absorbs its character depending upon the direction from which it travels; for example, in northern Europe and the British Isles, it is dry and gives us fine, clear, sunny weather, but when it travels south to the Mediterranean it absorbs modifications, becoming moist and unstable, so that southern Europe experiences showery weather with the possibility of thunder.

Areas of the globe other than the British Isles experience certain other types of air mass which bring important changes in their weather. For example, there is a high altitude air mass which is very dry and warm, and originates just above the Troposphere. This is known as 'superior air' and in summer it sometimes drops to Earth level, bringing exceptionally hot condition to the mountains and plateaux of the south-western states of North America.

Within the equatorial belt, the Equatorial air mass ensures typical tropical weather, while the area from Korea to India is affected by the Monsoon air mass.

This originates over southern Asia, absorbing the moisture-charged hot air from the Pacific Ocean. In winter, the air flows in the direction from Korea to India, deflecting back into the Indian Ocean the winds from the Pacific, and causes storms when the Maritime air mass travels into the central land areas to meet the dry tropical Continental air which battles for superiority.

In summer, the wind-flow is reversed and travels from India to Korea, producing torrential rain.

Generally speaking, that air mass which moves largely across expanses of water is heavily charged with the water vapour which has evaporated into it during its passage across the seas, and it is for this reason that the air is called Maritime.

The air mass which has travelled largely over continents will not be so heavily charged with vapour and is therefore known as Continental air.

In relation to our World Atlas, we can easily give a home, or source of origin, to the principal 'weather factories' responsible for the basic air masses which affect our experience of the weather; for example, there is an extensive zone of settled weather air mass between the European and African coasts on the east and the American coasts on the west, so that our south-westerly view of this zone is in pleasant anticipation of settled conditions and good weather prospects.

Away to the north-west, over Greenland, is another immense zone of generally settled weather, which in summer provides us with a cool spell of weather, and in winter sends us the 'seasonable' frosty days and nights that are often regarded as being bracing and healthy.

These great zones of generally settled weather which arise from the Greenland-African-European-American weather factories can be roughly merged together to form a broad Y-shaped area with its tail jutting out into the Pacific, its left arm embracing most of Greenland and its right arm lying across the Atlantic, North Africa and well down the Mediterranean.

It now becomes apparent why our British weather is so difficult to cope with, for the British Isles just misses being included in the European fine weather belt and falls instead within the boundaries of a vast zone of generally changeable weather which includes the North Atlantic on the west, southern Greenland on the north, and across Scandinavia to Siberia in the north-east (Fig 7). Here the bleak east winds originate,

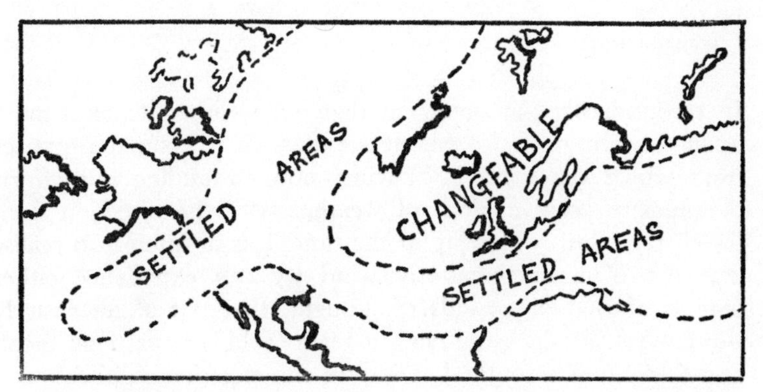

Fig. 7 *Weather belts of settled and changeable areas.* Note how the British Isles just misses an area of settled weather.

bringing us in winter our most severe weather and icy spells, while in summer, when the east winds blow from Scandinavia, we experience fine hot weather on dry hot air.

It is, therefore, now obvious that it is not so much the winds themselves which determine the type of weather we have, but the character of the air mass responsible for them, that the responsible air mass has formed its main character at its source of origin, and that this source may be hundreds of miles from the area experiencing the weather it brings.

The winds themselves are produced by the movement of the main air masses from areas of high pressure to areas of low pressure.

In the northern hemisphere, the winds in a low pressure area are caused by atmospheric suction to swirl inwards in a counter-clockwise direction, which is called 'cyclonic'. In the southern hemisphere, the winds are in a clockwise direction.

The winds in a high pressure area are caused to travel in a clockwise direction in the northern hemisphere, and are called 'anticyclonic'. They are counter-clockwise in the southern hemisphere.

For general convenience, the low pressure area is called a *depression* and the high pressure area is called an *anticyclone,* and as we shall see in the next chapter, these have their clouds, their weather and their winds.

Circulation

In reconsidering the statement that our weather depends more upon the *origin* of the parent air mass than upon the quarter from which the wind is blowing, notwithstanding our efforts to associate positive kinds of weather with the direction from which the wind is blowing at the time, it is of interest to realise that out of a north-easterly wind we can experience either periods of rain or periods of drought, that out of an easterly wind we can have either extreme cold or extreme heat, and that out of a southerly wind we can experience either warm sunny days or cold rain. All of which show that there is no simple relationship between weather and wind direction without some qualification as to the source and season – which, indeed, we shall attempt later in this chapter.

If the planet were still instead of spinning from west to east, the winds would travel with little modification from one place to another by virtue of the differences in pressure of their parent air masses. As it is, the spinning globe is pulling the atmosphere along with it, and so the winds are deflected from their path, the westerly winds being thrown outwards and southwards so that in the south they build up high pressure systems, leaving low pressure systems in the north.

The high pressure area is farther south in the winter and is found spanning the distance from the Azores to Bermuda, but in summer it travels north and across the British Isles, giving us, perhaps, several weeks of fine weather.

Around the Equator, where the heat is greatest, there is always a flow of warmed air rising from the Earth's surface, to be replaced by cooler air which rushes in beneath it to restore the natural balance.

The rising air continues to elevate itself, flowing in opposite directions to the ground wind, until it has cooled and descends again at a distant point by virtue of its weight. It circulates continuously in this manner.

At the North Pole the cold air is flowing away from the surface, and is made to circulate by virtue of the higher winds which blow towards the Pole. Between the two circulations, that from the Equator and that from the Pole, there is a third circulation originating at about latitude 60° degrees. This circulation is formed by the cold air from the north meeting the warmer air from the south. Exactly the opposite occurs in the southern hemisphere.

All winds, then, are seen to be caused by difference of atmospheric pressure, and blow from regions of higher pressure to regions of lower pressure (Fig 8).

These differences in pressure and the resulting winds are due to changes either in humidity or in temperature. For example, if one region becomes of higher temperature than a neighbouring region, the warm air from it will rise, owing to its lightness, and will pour in on the cooler region, while the cool air from this region will replace the air which has risen. In this manner two currents are formed :

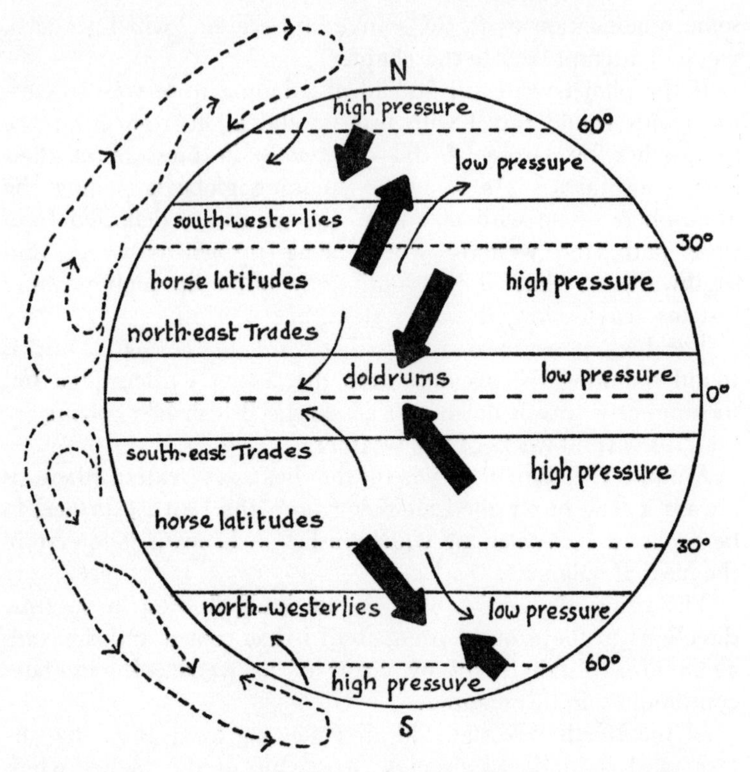

Fig. 8 *Pressure belts and circulation.* Winds cannot travel direct from area to area but due to the rotation of the Earth from west to east, are deflected, to the right in the northern hemisphere, and to the left in the southern.

a. travelling from the cool to the warm region on the *surface,* and

b. travelling from the warm region to the cool, *above it,* and this will continue until an equal balance is arrived at once again.

Land and sea breezes, as discussed shortly, are good examples of this process.

At the Equator, the weather is hot and humid, giving rise to rain, thunderstorms and gales. It is normally calm between these times and is otherwise referred to as the Doldrums, where ships under sail would frequently be becalmed for several weeks.

The effect of the rotation of the Earth is to cause the basic wind circulation to spiral, or slip, across the surface instead of travelling directly to areas of low pressure from areas of high pressure, as would be the case if the Earth were stationary.

This rotation causes anybody in the northern hemisphere to be spiralled to the right, and to the left in the southern hemisphere. The actual degree of deflection is variable, depending upon the distance from the Equator, since objects at the Equator move at about 17 miles a minute, those at latitude 60° at about 8½ miles per minute and so progressively more slowly towards the Poles.

A wind approaching the British Isles from the south would be deflected to the right, appearing as a south-westerly.

A wind approaching from the north is deflected to the left below the Equator and appears as a north-westerly.

Prevailing winds

The winds over land are changeable from day to day and from hour to hour depending upon the general appearance of world weather, and they cannot, therefore, be charted for reference. It is only over large water areas, where conditions are more stable, that the wind has a chance to take some characteristic course as a result of the basic circulation of the Earth's air masses.

These courses are called the 'prevailing winds', and are of great service to navigation as they represent at least one known factor on an often unpredictable journey. Old-time mariners made capital use of these known winds, and Christopher Columbus seems to have been an expert in the advantages they offered to supplement his navigation. The famous Kon-Tiki expedition was carried out entirely by the movement of the sea currents which have prevailed on that route for centuries, and which old-world seafarers used with confidence and precision.

One famous prevailing wind is the Gulf Stream, which sets up a current on the surface of the Atlantic northwards from the Gulf of Mexico and forms the warm south-westerly wind which we experience in the British Isles.

The pressure systems which produce areas of high and low

pressure are responsible for all the chartable winds of the world, and among these are the important Trade Winds and Westerly Winds.

The Trades blow in the area between approximately 30° and 10° North and South latitudes. Their average direction is from north-east in the northern hemisphere and south-east in the southern hemisphere. They are principally contributed to by the air that descends from the high-pressure sub-tropical areas, and are reasonably constant in their direction and behaviour, so that it is possible to anticipate what kind of winds are to be expected from day to day. The Doldrums is the area where the Trade Winds meet.

On the other hand, the Westerlies arise from a complex system of irregular pressure areas, and are therefore of a variable nature, their paths lying to the north and south of the calm regions. In the northern hemisphere, these winds are affected by other weather systems and in their tracks over land masses, but they blow almost unaffected in the southern seas and with predictable regularity as there are hardly any land masses to modify them.

In the North Atlantic and the North Pacific in latitudes 45°N, and in the southern hemisphere in latitude 45°S, there is a belt of stormy winds known as the Roaring Forties which arise from the Westerlies, and which, in the southern hemisphere, provide a belt of almost continuous storms.

The Westerlies of the British Isles flow eastwards from the calm high-pressure areas known as the Horse Latitudes just 30°N and S of the Equator.

Winds and air currents

> 'For raging winds blow up incessant showers;
> And when the rage allays, the rain begins.'
>
> *Shakespeare* (*Henry VI*)

Wind is a horizontal movement of air which, in connection with flying, can modify the aircraft's course in relation to the ground, and in connection with sailing, can modify the ship's course in relation to a direct line between the point of departure and its destination, so that adjustments in navigation have to be made to ensure that they are not blown off course.

A vertical movement of air is called a current.

An oblique movement of air is both a wind and a current. That part which is wind is called the horizontal component, and that part which is the current is called the vertical component.

It is the vertical component of such a movement that the glider pilot seeks, for, depending upon its strength, he may reach great heights by soaring in the up-current. The aeroplane pilot, on the other hand, is affected in another way. When he flies through a column of rising air, his aeroplane lifts with it and then drops in the depression as it leaves the up-current behind it. This is the cause of the 'bumpiness' so often experienced during low-altitude flights.

Wind does not always travel at constant velocity. That which is near to the ground is readily redirected by obstructions such as buildings, trees and hills, and will vary perhaps between 25–45 miles per hour, and between west and south in direction, while extra violence is caused by hills and cliffs. This lack of uniformity is known as being gusty.

Land and sea breezes

> 'A wind generally sets from the sea to the land during the day, and from the land to the sea at night, especially in hot climates.'
>
> *J. F. Daniell*

Land and sea breezes are due to differences in temperature over land and water areas. When two adjacent areas become unstable by developing unequal temperatures, the air of the warmer region, being less dense than the colder air mass, will rise and flow in from above, while the colder and more dense air below will flow outwards to replace the ascending warmer air. In this way two currents are formed, one flowing from the colder to the warmer region across the Earth's surface, and one flowing from the warmer to the colder region above it in an attempt to restore the stability of the adjacent areas.

This kind of movement, which produces wind, gives rise to

what we call 'local' winds, of which those produced by the temperature differentials between land and sea masses are at one time or another experienced by most of us and are therefore readily understandable.

During the daytime, when the land becomes warmer and its pressure is lower than that of the water, the sea or lake breeze is created and blows inwards across the land, beginning as a gentle movement and increasing in strength in the heat of the day (Fig 9).

At night, when the land becomes cooler and the area of water becomes warmer by comparison and is in a state of low pressure, the land breeze blows seawards into the night, fad-

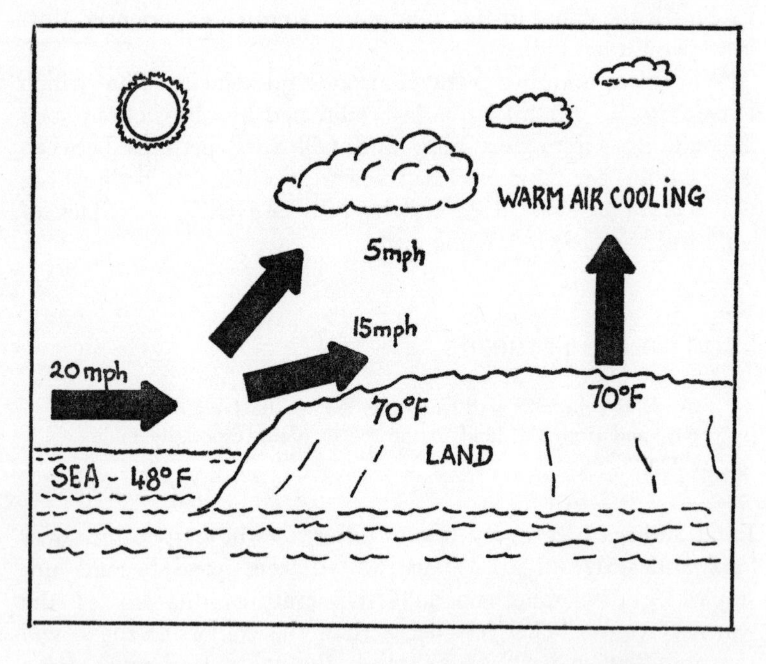

Fig. 9 *The sea breeze*. By day the Sun heats the land more than the sea. As the heated air rises it expands and the cooler air flows in from the sea to replace it, becoming strongest at about 3 o'clock in the afternoon. The sea breeze is at once slowed down as it strikes the surface and other obstructions. Some of the air which rises reaches condensation level and forms cumulus cloud.

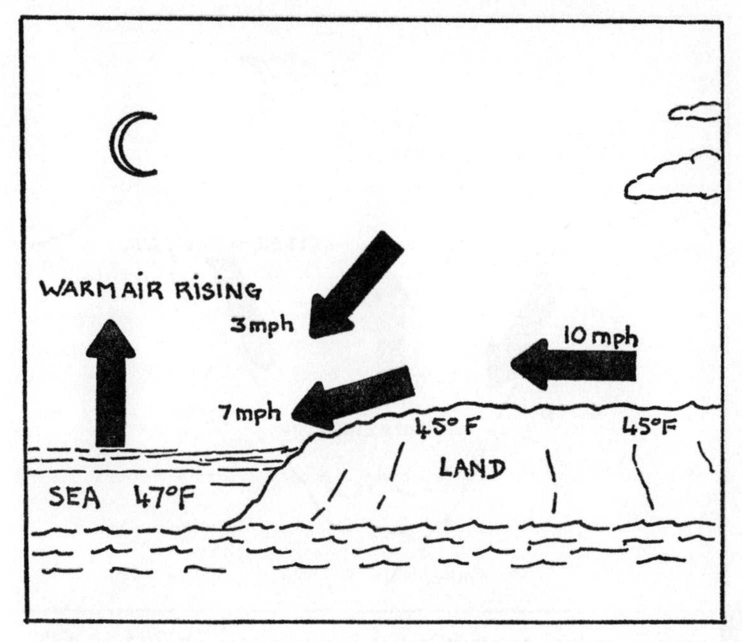

Fig. 10 *The land breeze.* At night, the land will cool more rapidly than the sea area, and this reverses the conditions of the daytime. The air over the sea is lighter and warmer than that over the land, and consequently rises from the surface, being replaced below by the flow of cold air from the land, and this continues until conditions are stable once again.

ing away in the morning in favour of the sea breeze (Fig 10).

The old time fishing industry for centuries depended upon the effect of these breezes for outward and return journeys of its fishing fleets. The boats set sail at night before the strong land breeze, and returned on the sea breeze at noon.

Mountain winds

These are due to cooler air over mountains graduating *down* into the warmer valley at night to replace the warmer air which is being radiated from the ground, and they are known as Katabatic winds (Fig 11).

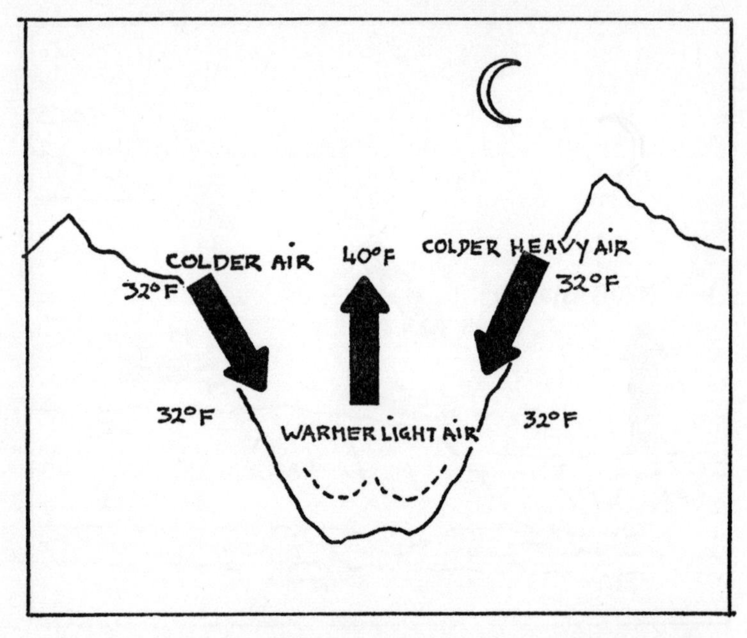

Fig. 11 *Katabatic wind.* At night, the hills radiate into the air more heat than the valley between them. The air contacting the hill-tops becomes colder and dense, and begins to sink downwards as the warm air in the valley rises. Mist often forms and the wind carries it down into the valley during the night. The movement of air so created is known as a katabatic wind.

Valley breezes

These blow *up* the valley when the ground becomes warmer due to the heating effect of the Sun, and they are called Anabatic winds (Fig 12).

Orography

'They are wet with the showers of the mountains.'

Job xxiv, 8

When the wind blows against a hill or mountain, it is forced upwards into cooler regions where condensation takes place and

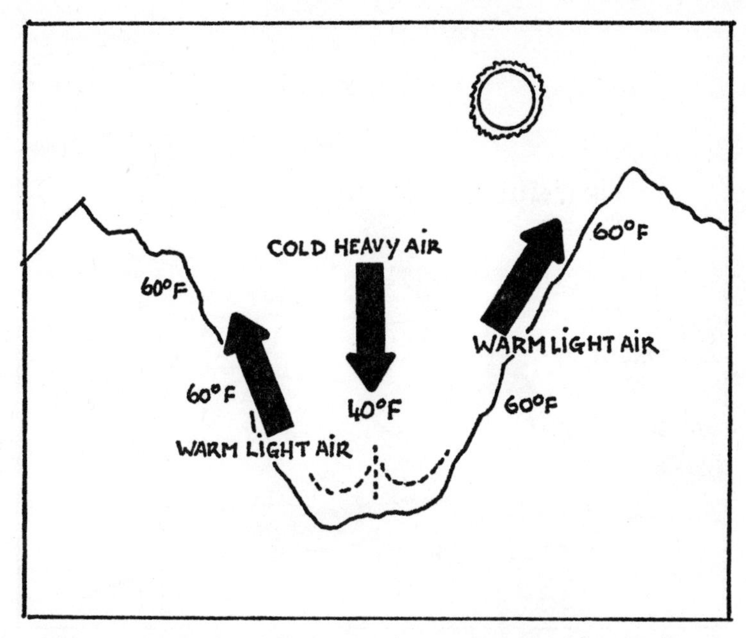

Fig. 12 *Anabatic wind.* As the slopes are heated by the Sun, the air temperature rises and the air expands and flows slowly up the slopes, being replaced by the cooler air at the centre. The movement of air so created is known as an anabatic wind.

forms banks of cloud at the top of the hill. This is called an orographic wind. The cumulus cloud that it forms is called orographic cloud, and is the result of convection since it is rising air which causes it. Rain which falls from these clouds is called orographic rain (Fig 13).

Such conditions are typical of a range of hills or mountains, when the wind, in striking the oblique surface, is deflected upwards for some considerable distance and in rising is cooled and forms cloud. The glider pilot seeking out these rising air currents will identify this type of cloud and make for the rising column of air which causes it so that he may use it to gain height. This is called 'slope lift'.

After passing over a hill, the air often develops an oscillation as it attempts to resume a straight course and, in doing so, produces regularly-spaced clouds parallel to the hill. These are

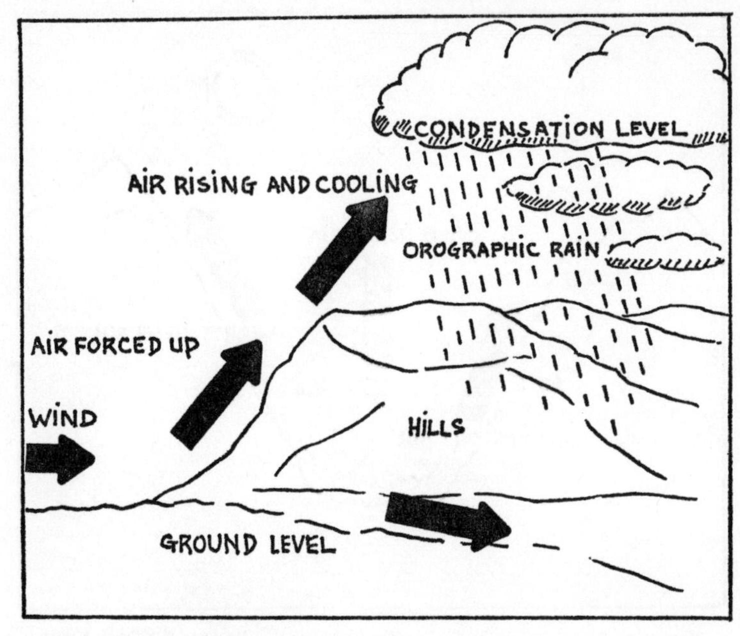

Fig. 13 *Orography*. Wind, forced against hills and mountains, rises and cools, and at condensation level forms banks of cloud. This movement of air is called an orographic wind. The cumulus cloud so formed is called orographic cloud. The rain which falls from such cloud is called orographic rain.

called lee waves, and light aircraft caught in the down-draught following the upward movement which produces them may be brought down or buffeted' in the oscillation of the air movement.

Thermals

The Sun, when powerful enough, penetrates the dry air and strikes the ground, and as the ground temperature rises it radiates heat waves which rise into the surrounding air. The air itself then starts to rise as it becomes heated. As the air ascends it becomes cooler and is replaced by more warm air from below so that there is a continuously rising column of warm air which persists for as long as the conditions causing it are maintained, and this is called a 'thermal'.

The strength of a thermal is related to the kind of ground which gives rise to it. For example, it may be expected more over dry, ploughed land than over rough scrub, and more over empty fields than over green meadows, as the former absorbs more heat.

An indication of a thermal source is a white cumulus cloud with a flat hollow base, and glider pilots seek out these strong thermals to gain lift for increasing their height. They can in fact fly for many miles by using a succession of developing thermals.

Since a thermal is essentially a rising column of warmed air, its production is not necessarily dependent upon the Sun as being its original heat source, since any artificially created heat source will have an identical effect providing it is of sufficient strength; for example, convection cloud is often seen above power stations, where the generated heat rises in a column up to the condensation level of the surrounding air.

Usually, cumulus clouds are caused by several thermals which become mixed with one another to form a large cloud or series of clouds. Sometimes the rising thermals reach a temperature inversion, or stable layer, which resists further development, but when the air below the inversion is raised above its condensation point, the thermal produces cloud which presses through the inversion layer like a large bubble. This sometimes produces the smooth dome-shaped cap of cloud which we call pileus and which was first mentioned during our study of the clouds.

When the thermal is not strong enough to impress the inversion, the clouds formed are then spread out below the inversion like steam from a boiling kettle on reaching the kitchen ceiling.

Friction

Wind, as with all other elements and substances, is affected by friction; that is to say, the wind velocity at the Earth's surface is slowed down by contact with the rough and irregular surface until there is no movement at all from the particles of air in contact with ground. This effect becomes less with increasing distance from the ground up to a height of about 2000 feet, at which height it is called the gradient wind, and is considered to be beyond the direct influence of surface friction.

Wind changes, veering and backing

Wind is not constant in direction, and changes possibly from hour to hour. Change in a clockwise direction is called 'veering' and change in an anti-clockwise direction is said to be 'backing'. A change of direction may also mean a change of velocity.

A veering wind may indicate a change for the better, while a backing wind usually brings a change for the worse, since the wind backs, or shifts against the Sun, as a depression advances, and veers, or shifts with the Sun, as the depression passes to the east with a following anticyclone coming in from the west.

Some of the old weather observers have left us their words of wisdom on this very subject:

'The veering of the Wind with the Sun, or, as sailors say, right-handed, prognosticates drier or better weather; the backing of the Wind against the Sun, or left-handed, shifting, indicated rain, or more wind, or both.' *Fitzroy*

'If in unsettled weather, the wind veers from SW to W or NW at Sunset, expect fine weather for a day or two.' *Fitzroy*

'If the wind veers from N to NE in Winter, intense cold follows.' *Dove*

'If the wind in daytime shifts from N to SW, or S, rain is pretty sure to follow, if on the other hand, it shifts from S or SW to N, the weather will probably clear up.' *from Devon*

Violent winds

A *gale* is accepted as being a wind with a velocity of about 38 miles per hour, at which speed one may expect to see the trees in general movement, possibly with twigs breaking off.

A *fresh gale,* between 39 and 46 miles per hour, removes twigs and causes high waves at sea, producing widespread foam.

A *strong gale,* between 47 and 54 miles per hour, may blow off chimney pots, television aerials and damage any lightly constructed building. At sea, the waves are high and dense foam drives before the wind.

A *whole gale* blows at 55–63 miles per hour, and will uproot trees and damage houses. The sea is turbulent with high waves and large areas of foam.

Storms at 64–75 miles per hour cause widespread damage and, like the *hurricane* at above 75 miles per hour, are fortunately rare occurrences.

The same character of the hurricane is called a *typhoon* in the China seas, and a *cyclone* in the Indian Ocean. The storms are furious with whirling winds at the centre at speeds of 100–200 miles per hour at times. At the 'eye' of the storm, that is the dead centre, winds may cease altogether or become mild.

When the hurricane reaches the sea it is a great menace to shipping, causing mountains of water to rise into the air. Normally, as it moves inland, the storm dies out owing to the lack of water vapour which has been feeding it whilst over the sea.

East of the Rocky Mountains in the United States, the greatest menace is the devastating *tornado,* which is heralded by strange colours among the clouds. Following this, a towering column of water vapour drops from the sky, only a few yards wide but roaring and whirling like a demon as it destroys houses and life in its path. It travels at about 30 miles per hour, but the winds at its centre are around the 200–mile per hour figure.

In New South Wales, a hot northerly wind called the *Brickfielder* precedes a cold wind known as the *Southerly Buster.*

The *Berg* is the name given to a hot wind in South Africa, which dislodges the heat from the high plateau and causes it to flow downwards to lower regions, where, in spite of the discomfort it causes to human and animals, it accelerates the ripening of the fruit in the valleys.

The tornado-type of wind which produces torrential rain and brings extensive damage to houses and crops in Australia is known as the *Willy Willy.*

Hot, dry, sand-ladened winds blow furiously across the deserts where they draw up clouds of sand and make the great sandstorms which have always been one of the hazards of desert travel. In Egypt, the wind is known as the *Khamsin,* while in the Sahara it is called the *Sirocco.* The *Haboob* in the Sudan is a gigantic dust spiral which reaches up to 5000 feet or more, and covers everything in its path with choking dust.

Another wind which drives across the Middle Eastern deserts, is the *Harmattan,* which has been known to carry dust and sand far out to sea, forming a dense fog.

The *Föhn* wind blows for several days in succession and brings extremely dry heat to the Alpine valleys. On the other hand, the French Alps provide the source region for the bitter hurricane-force wind that races through the Rhône valley. This is known as the *Mistral* and, although not normally accompanied by storms or rain, will bear down with full force on bright, sunny days.

From the same source as the Mistral, the *Bora* produces similar patterns of behaviour in the Adriatic Sea.

This does not exhaust the list of 'local' winds by any means, since there is the south-east *Solana* of Spain, the *Pampero* which blows from the Andes across Buenos Aires to the sea, the cold September *Northers* on the Gulf of Mexico, and the *Levanter* in the Mediterranean; but there is little point in pursuing the matter further at this stage, except, perhaps, to be thankful that the British Isles is not in the path of these peculiar winds, and that our most uncomfortable experience is the east wind which prevails in the spring, and which is always dry, with low humidity, and brings a good deal of personal discomfort to the ailing and elderly.

As with all other winds, gales are caused by the difference of pressure and the resulting attempts to equalise by the replacement of air from one area to another. The greater the difference, the greater the volume of air to be moved, and the greater the general disturbance of the atmosphere around us.

Gale warnings: North and South Cone Indicators

Certain ports, fishing stations and all air stations are warned of approaching gales. The gale is indicated visually at the station by the hoisting of a black canvas cone which is displayed until the gale, or the danger of a gale, has passed.

The term North Cone is given when the point of the cone is uppermost. The term South Cone is given when the point of the cone in inverted. Thus:

North Cone ▲ South Cone ▼

The North Cone is hoisted for gales beginning in a northerly direction, or if a gale originating from east or west is expected to change to a northerly direction.

The South Cone is hoisted for gales beginning in a southerly direction, or if gales originating from east or west are expected to change to a southerly direction.

Gale warnings are also broadcast in general weather bulletins and on special radio frequencies which concern aircraft and shipping lanes, but before these facilities of modern times, the hoisting of cones was the only warning available to the inshore sailor, while others, out of sight of the coastal stations, relied upon their ability to interpret their barometer and storm glass.

Wind force

The wind force is estimated and expressed according to a scale laid down by Admiral Beaufort as long ago as 1805, and which has since been know as the Beaufort Scale.

As the wind is slowed down considerably owing to friction as it draws closer to the ground, all observations are based on the speed at 33 feet above the ground, made in a position unlikely to be influenced by obstruction or deflections from nearby buildings.

The speeds are average between certain limits as shown in the table below, and the strength observed is transferred to the weather map in the form of an arrow with a number of full-length or half-length feathers. A full feather represents 10 knots. A half feather represents 5 knots, so that one and a half feathers on a shaft represents a total of 15 knots of wind speed, and the front end of the shaft points the same way as the wind is blowing.

No.	*Wind*	*Arrow*	*Speed mph*	*Effects*
0	Calm		Less than 1	Smoke rises vertically.
1	Light air		1–3	Direction given by smoke but not by wind vane.
2	Light breeze		4–7	Vane moves. Leaves rustle. Wind felt on face.
3	Gentle breeze		8–12	Leaves and small twigs in movement. Light flags extend.
4	Moderate breeze		13–18	Small branches move. Dust swirls along with light paper.
5	Fresh breeze		19–24	Small trees sway. Crests appear on inland water waves.
6	Strong breeze		25–31	Large trees' branches sway. Telegraph wires hum.
7	Moderate gale		32–38	Walking difficult. Whole trees move. Dust flies high.
8	Fresh gale		39–46	Walking more difficult. Twigs break off. Windows rattle.
9	Strong gale		47–54	Loose bricks and chimney pots dislodge. General damage.
10	Whole gale		55–63	Trees uproot.
11	Storm		64–72	Widespread damage.
12	Hurricane		72-plus	Anything may happen.

The Wind Vane

'Every wind has its weather.'

Bacon

The points concerning wind which we notice immediately are its strength, its temperature, its humidity and its direction.

We are reminded of its strength by the disturbances it may cause. Some winds are not strong enough to move the wind vane but may be indicated by blowing smoke from chimneys along with it. At other times, the strength may be enough to rustle trees or to make the washing on the line flap about furiously.

Dust may be driven from the streets and into our faces, clothes may be displaced, and we may have to lean into the wind as it seems to try its best to dislodge us from our feet.

On a rainy, windy day, umbrellas blow inside-out, while in the teeth of any strong wind, hats sail through the air and people are almost blown over as they cross the street away from the protection of the surrounding buildings. These occurrences make us aware of its strength.

During the summer we may seek a welcome place for a cool breeze to relieve the heat. A bather, on the other hand, may find himself suddenly chilled when emerging from the warm sea.

During the winter and in bad weather, the wind may blow with icy fingers about our ears and encourage us to don warmer clothing. These occurrences make us aware of the temperature of the wind.

The amount of dampness felt on the wind is an indication of the humidity of the air which carries it.

Sometimes we notice that the wind will blow chimney smoke towards a land-mark which we are able to see from a convenient window; at other times the trees to our left will be bowing before the wind and away from that same land-mark. Sometimes the wind will smash with force against the windows in the front of the house, and on other occasions it will vent its strength against the windows at the back. These occurrences make us aware of its direction.

The instrument used to indicate this direction is the wind vane, otherwise called the weather vane or the weather cock, the latter being after the practice of mounting it with a model cockerel, although today it is more usual to use an arrow.

The wind vane should be mounted on a high point on a building or on a pole between 30 and 40 feet above the ground and well away from any obstruction which might influence the direction of the wind.

The four cardinal points of the compass, north, south, east and west, are extended on arms at the top of the mounting, north being lined-up directly on the *true* north. Above this, and mounted so that it may rotate freely, is an arrow with a large tail. As the wind strikes it, the arrow will take up a position heading into the wind, so that by regarding the arrow head and checking it with the point of the compass to which it is nearest, one may obtain a reading of the wind direction (Plate 22).

A north wind is one which blows *from* the north, and a south wind is one which blows *from* the south, so that direction is given as the point *from* which the wind is blowing.

Over a period it is possible to decide if the wind is veering or backing. Should the change be from north-west to west or south-west, it is backing and may bring poor weather. If the change is from south-west to west or north-west, it is veering and may bring fine weather. Naturally, any indications given by the direction of the wind would have to be confirmed by reference to the other weather instruments in use.

It is important that we should know what type of wind we are experiencing, since, as we have seen throughout this chapter, the direction from which any air mass is travelling has a bearing on the weather it brings.

It will not come amiss at this point to mention how the 'weather cock' got its name. You will remember at the time of the Last Supper according to St Mark, chapter 14–30:
'And Jesus saith unto him [Peter], Verily I say unto thee, That this day, even in this night, before the cock crow twice, thou shalt deny me thrice.'

It was to commemorate this dramatic prediction that Pope Nicholas I, in the middle of the ninth century, commanded that the highest steeple or pinnacle of every cathedral, abbey and parish church throughout Christendom should be surmounted by the figure of a cockerel.

Whether this was wholly adhered to we have no means of knowing, but we can be sure that the practice was fairly widespread, and that since the wind vane came into use and was normally mounted on the highest point of a building, the cockerel was awarded pride of place above all else.

As we must all have observed, there are now many interesting deviations from the original cockerel, and all sorts of patterns and figures may appear in association with the direction-indicating arrow.

The Anemometer

A light wind may bowl pieces of paper and leaves along the ground for short distances. It may be noticed that a certain screwed-up piece of paper passes 4 gateways in about 4 seconds, it rests, then along comes the wind again. This time it passes 5 gateways in 4 seconds, and after a pause it takes only 3 seconds to pass 5 gateways. The simple conclusion is that the wind is becoming stronger and is increasing in speed, but such simple visual indications are not nearly accurate enough for the meteorologist, and it requires the aid of a special instrument called an Anemometer to record for us the speed of the wind about us.

Leonardo da Vinci suggested the first anemometer for measuring the strength of the wind by its action in depressing a metal plate, the base of which could be related to a graduated scale on its supporting bracket.

The pressure type of anemometer was introduced into the science of weather recording by Lind in 1775, and it was not until 1843 that T. R. Robinson, FRS introduced the first cup anemometer, which gave the wind velocity independently of the wind direction, for which a separate vane was necessary.

At that time, it was thought that 4 hemispherical cups mounted on the horizontal arms of a spindle gave the best results, and the spindle was geared to a recording dial which measured the distance travelled by the cups and consequently converted this to the speed of the wind which drove the cups around.

It was not until 1926 that J. Patterson disproved the original theory and reduced the modern anemometer to 3 cups only, so that as the cups spin around in the modern recorder, their pivot operates an electrical generator in the base of the unit, and electrical impulses are transmitted through wires to recording dials some distance away. The dials read in knots or miles per hour (Plate 23).

Upper winds

We have seen in Chapter I how we use the radio sonde to bring us information concerning the upper atmosphere, and how its path can be tracked by radar equipment picking up signals which are reflected from its special reflector.

It is, however, expensive on the one hand and not always

practicable on the other to send up this kind of balloon with its sophisticated radio equipment. And so weather observers prepare and release a Pilot Balloon which is without parachute and radar and is used solely to determine the wind speed and direction. Its path is followed by a meteorological assistant using a pilot balloon theodolite, which is a surveying instrument for measuring horizontal and vertical angles, and is thus used to determine the behaviour of a balloon in free flight.

At night, a paper candle lantern is attached to the balloon so that it may be seen during its ascent.

In order to convert the readings obtained from following the balloon with a theodolite into wind speeds and directions, trigonometrical calculations are necessary. These can be done on a polar diagram or by means of a meteorological slide rule, the details of which we need not concern ourselves with in this book.

More to the point, it is time to come down to Earth and find out what the various winds have in store for us when they come our way.

What the winds foretell

> 'Dost thou know the balancing of the clouds, the wonderous works of Him which is perfection in knowledge? How their garments are warm, when he quieteth the earth by the south wind?'
>
> *Job xxxvii, 16, 17*

We have devoted considerable time to detailing the characteristics and behaviour of the winds, for like the clouds which ride before them, they are of prime importance in the intricacies of our weather systems.

It must be borne in mind, as with all weather observations, that no single item can give an accurate pointer to the weather; it must be derived from a combination of several factors such as cloud, temperature, humidity, pressure and wind.

South wind

This precedes a depression and is comparatively mild. In summer it may bring heat, but to us, generally humid and thundery weather with poor visibility.

South-west wind

This is the prevailing wind of the British Isles, and approaches having travelled far with little variation in direction, bringing moist and stable Tropical Maritime air. When it arrives on a really hot day, large amounts of cumulus are created.

In winter this may grow to cumulonimbus and spread rain. In summer it may form nimbus unless it remains over land for a considerable time, but may bring squally weather. It is generally mild in winter and warm in summer.

West wind

This is cooler than the south-west wind and is usually dry. Following poor weather, it may point to a decrease of rainfall and clear visibility.

North-west wind

Wind from this direction indicates Polar Maritime air. It is unstable after a long passage over warm seas, and may carry the cumulonimbus. Thermals may be expected to start when the air meets land, and the humidity may be averaging around 70 per cent.

Cumulus may form at high levels. Passing Great Britain, it becomes more stable as it approaches the Continent. It is cool in summer and cold throughout the remainder of the year, and is usually dry with good visibility.

North wind

This is cold in winter and spring, but cool in summer. It is generally dry with good visibility.

North-east wind

The north-east wind is mainly responsible for the April showers which come from the fairly low cumulonimbus appearing in spring. The air approaches across the North Sea, starting off as Polar Continental air and assuming Maritime characteristics during its journey.

It is cold in winter and spring, and generally dry with a possibility of frost. If it appears where there is a depression travelling up the English Channel and crossing south England, it may bring heavy rain. In summer it may herald a heat-wave.

In winter, where there has been a high pressure in the north, and a fairly widespread severe frost, this wind will probably mean a dispersal of it. Further confirmation of this may be borne out by a falling barometer and an overcast sky with a yellow-green tint.

With the overcast sky, the wind may veer from north-east to north and north-west. As the wind shifts round to the west it is often accompanied by snow, sleet and rain in succession. Humidity will increase and the temperature rises.

The fine weather characteristic of the north-east wind is first shown by a high-pressure reading, rising or stationary, a hazy horizon and clear blue sky, with low humidity, and a cold to rising thermometer according to seasonal effects.

If the wind velocity remains steady and, if anything, veering towards the east or south-east, it would be safe to predict settled conditions.

East wind

This prevails in spring and is very dry. It has humidity values around 29, 37 and 44 in the Highlands.

Cold in winter and spring, biting and sharp, the east wind is regarded by many as a bringer of discomfort and ill-health.

In summer it is mainly cool to mild.

South-east wind

This generally appears as cold in winter and warm in summer. It is not uncommon for it to herald the approach of a depression, having originated as a south-westerly and circled the depression during its journey. In this case, the winter character is a warm spell, and in summer it becomes close and sticky.

In the summer season, the south-east wind may bring heat from the Continent. A thundery spell may be indicated by a fall in pressure as a cool south-westerly comes in above the south-easterly, most likely giving rise to altocumulus coming in from the south.

The *fine weather* south-east wind represents settled conditions, and is usually the outcome of polar currents arising from the north of Russia and eventually spreading over Europe and affecting the British Isles.

The barometer will read high owing to the high pressure descending upon it, and the dry air mass will produce rapid evaporation from the wet bulb thermometer.

The rough weather south-east wind is characteristically warm and moist so that it will affect both the wet and dry bulb thermometers. In consequence of the approaching storm, the barometer will be dropping and the sky indications will normally be clear and streaked with cirrus, followed in due course by the nimbus.

The wind blows stronger as it accompanies the storm from the west, veering to the south as the weather advances.

In winter, the temperature will be low and the polar air will spread ice on exposed water. In spring, it will normally be dry, while in summer and autumn we may expect a spell of fine, sunny weather.

After some tenacious personal observations it will become possible to augment these notes with local indications, but in the meantime, the following chart will serve as a ready reference guide to the winds.

Wind	*Winter*	*Summer*
S	Precedes a depression. Mild.	Often humid. May herald hot spell. Visibility poor. Haze.
SW	Bringing low pressure area centred to the NW. Mild and showery. Thaw after cold spell at about 43°F.	Warm, cloudy, showery, stable air. Much cumulus formation on hot day.
W	Approach of cooler period and decrease rainfall. Clear visibility.	Change to cool weather. Clear, dry.
NW	Cold. Generally dry. Frost at about 32°F. Visibility good. Centre of depression in NW.	Cool. Normally dry. Gives rise to cumulonimbus and thermals.
N	Cold in winter and spring. Generally dry with good visibility.	Cool. Dry. Visibility good.

Wind	*Winter*	*Summer*
NE	Depression passing followed by high pressure in W. Cold in winter and spring. April showers. May mean dispersal of frost after high pressure in north. Overcast sky veering to N and NW precipitation may accompany it.	Observe for warm spell. High pressure, hazy horizon, low humidity, rising barometer. Settled conditions if speed remains with tendency towards E or SE.
E	High pressure to N. Low pressure to S. Cold spells. Frost in spring. Generally dry with low humidity.	Mainly cool in the east and mostly mild in the west. Dry.
SE	May herald depression from W. Chill continental winds changing to mild southerly. May indicate warm spell. Low temperature and formation of ice on exposed water. Spring – dry.	In front of depression from the west, warm and sticky. Heat from Continent. Summer and Autumn expect fine, sunny spells.

We have now examined the behaviour and characteristics of the wind which arise from the air masses which form their main features at their individual source regions, and we have seen that the movement of these masses from one zone of the globe to another is caused by, and gives rise to, areas of high pressure and areas of low pressure which in turn produce their characteristic wind circulation. These are the depressions and anticyclones which predominate in our weather forecasts, and which we are about to examine in more detail in the following chapter.

6/OUR WEATHER SYSTEMS

'A veering wind, fair weather,
A backing wind, foul weather.'

Depressions and anticyclones are individual circulation systems which bring us our varying experiences of the weather as it occurs close to the Earth's surface. This circulation is not that which takes place between different vertical levels of the atmosphere, but that which takes place as the result of the flow of air from areas of high pressure (settled to fine weather) to areas of low pressure (rain to stormy) within the surface layer. Depending upon the direction from which these circulations appear, they have characteristics which are predominantly Polar or Continental, Maritime or Tropical, having been modified *en route,* not only by the land masses and sea areas over which they have travelled, but also by local variations caused by forests, mountains, lakes, valleys, coastal influences and season.

The winds which arise from the movement of air masses from one zone to another and from one pressure to another do not travel in straight lines direct from high- to low-pressure areas, but due to the rotation of the globe, they are deflected from a theoretically direct route and appear to travel on a curved path so that the depression is recognised by the fact that is has a wind rotating anti-clockwise and inwards to the centre and brings unsettled weather; whereas the anticyclone has winds which rotate clockwise to a high-pressure centre, and brings fair to fine weather, although it must not be assumed that conditions in an anticyclone are always bright and sunny.

The barometer, measuring the pressure of the atmosphere, reads low at the centre of the depression, perhaps in the region of 29¼ inches (990 millibars), and reads high, around 1034 millibars in the anticyclone, for which reason the depression is frequently referred to as a 'low' and an anticyclone is referred to as a 'high'; thus confirming the pressure of the type of air mass which is moving-in overhead.

Since the advent of the weather satellites which now orbit our Earth, it has been possible to study the exact formation of weather systems over the earth's surface and to confirm the cloud and wind patterns which contribute to the character of depression and anticyclones, and at individually selected areas on the globe.

It is fascinating to see on an actual photograph of a depression how the wind systems are whirling the cloud formations around and inwards in the typical anti-clockwise direction, and how surrounding areas can be recognised as familiar troughs, cols and fronts (Plate 24).

Pictures of the Earth, taken from some 22,300 miles in space, show clearly the outlines of the continents, although they are often heavily distressed by cloud and weather formations. Nevertheless, professional forecasters and climatologists are able to study these remarkable pictures with a view to developing more accurate techniques of forecasting, especially as frequent radio transmissions enable the movement of weather systems to be observed by the satellite and received at its Earth station with the minimum of delay, thus eliminating the tedious plotting of the hemisphere prior to discovering that a storm area is on the move (Plate 25).

For convenience, and since it has bearing on our later discussions about weather maps, we may represent the weather systems by means of schematic diagrams which show how the winds revolve around the pressure centre in either clockwise or anti-clockwise directions, depending upon whether it is a centre of high or low pressure respectively (Fig 14).

The behaviour of a depression

The beginning of a depression is when two vastly different air masses lie alongside each other at a division known as the *polar front*. This consists of a mass of polar air from the north-east running on the north side of the division, and subtropical air from the south-west running on the south side of the division – two masses of entirely different air proceeding in opposite directions, both battling for superiority along the line which divides them.

The polar air masses may, on the other hand, come from the west or south-west, having described a curve around a depression to the north, so causing the conditions whereby both

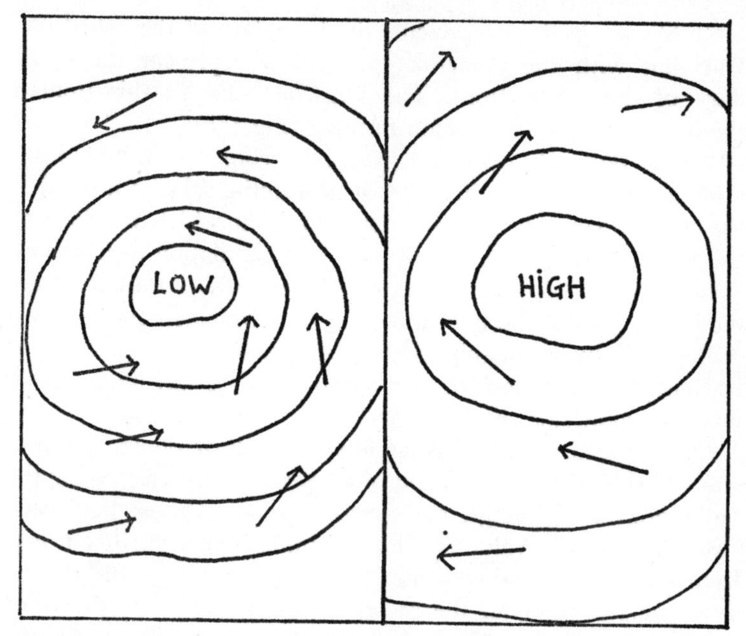

Fig. 14 Schematic representation of wind direction in low and high pressure areas.

air masses at the time of contact are travelling together and parallel, with the warmer air mass having the greater velocity.

For the purposes of our theoretical depression we shall have our two air masses now flowing in opposite directions along the dividing line which is called the front, and this may continue for some time during which each mass is struggling against the other – the cold trying to slide in under the warm, and the warm trying to rise above the cold (Fig 15a).

This combat may diminish into nothing more than a fresh breeze with neither side having established an advantage; alternatively, a front may be set up which is the beginning of a depression.

This front occurs when the division between the parallel air masses becomes distorted and forms a kink or bulge in its continuity. This is the moment during which the centre of the depression is borne.

In a schematic diagram of this situation, the line with the

tooth-like projections represents the boundary of the cold front, and that with the dome-like projections represent the boundary of the warm front. The kink or bulge in this front is the first sign of the depression forming (Fig 15b).

As the bulge develops, the warm air rises over the cold air, and the cold air flows in under it and anti-clockwise around to the rear to connect up with the cold front which is following the warm front round. In this way the circulation of a depression is formed, and the centre of it is at the tip formed by the narrowest part of the two fronts, an area which is schematically represented by a roughly circular line with the word 'low' at its centre, and a note of the barometric pressure in millibars or inches alongside (Fig 15c).

Outside and around this circle we can draw others to represent the theoretical connection of all weather stations having the same pressure readings, and from this we can see how each circle increases in value by about 2 millibars the further it gets from the 'low' reading centre. So the first circle outside the centre one reads 998, the second circle reads 1000, the third 1002, and so on until, perhaps, the pressure readings continue to rise as the circles progress outwards into other weather systems, and we find them merging into the high-pressure lines of an adjacent anticyclone.

As the depression deepens, the frontal lines become longer, thus forming a curving inverted V, with the cold front following down behind the tooth-edged line and the warm front running before the dome-edged line; the space between them in the inverted V is known as the *warm sector*. These fronts are given separate treatment shortly, so that their characteristics may be more fully understood.

These areas of low pressure are responsible in the main for the variable weather we experience, and it is possible for a succession of depressions to approach and cover the British Isles for several weeks in succession with little variation, giving us periods of bright weather followed by low cloud, overcast skies and rain.

A depression includes two velocities and direction. The first velocity is the anti-clockwise wind of the depression's own system, and the second velocity is the movement of the whole system across the globe; the direction in which it is travelling

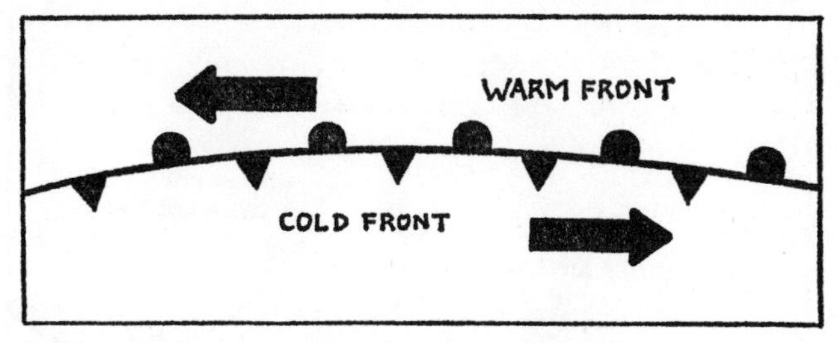

Fig. 15a *The meeting of the fronts.* Two air masses travelling in opposite directions meet at the polar front.

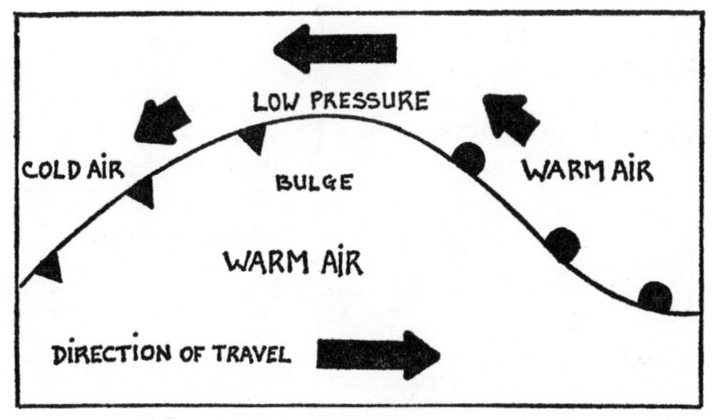

Fig. 15b *The bulge.* A bulge, or wave, forms which commences the wind circulation

is naturally another important factor, since we are then able to judge, by knowing its approximate speed, when it is likely to be affecting any particular point in its path.

The average wind in the system is about 30 miles per hour, which is regarded as a strong breeze, giving rise to whistling in the telegraph wires and the disturbance of tree branches and loose earth on exposed land areas. At sea, however, winds may be up to gale force and producing large waves of possible danger to small craft.

The system continues to work in this way until the depression declines and dies out by slowly filling in with the cold front narrowing the warm sector and overtaking the warm front, the process of which is called an *occlusion* (Fig 15d).

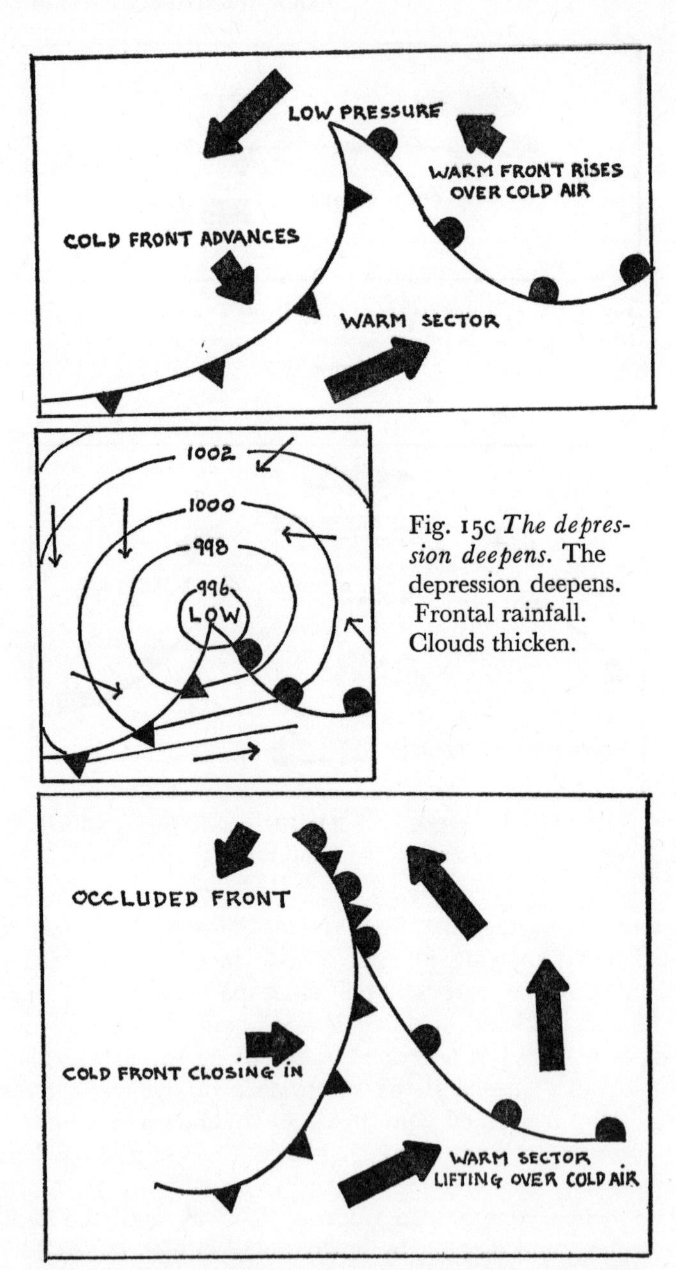

Fig. 15c *The depression deepens.* The depression deepens. Frontal rainfall. Clouds thicken.

Fig. 15d *The occlusion.* When the cold front overtakes the warm front the depression declines and is occluded.

During this time winds and rain are still to be expected, and it may take several days to complete, although the completion may not be sharply noticeable to the casual observer as it is likely to be followed closely by another depression which has been forming in the oncoming air mass.

The size of a depression may vary from a few hundred miles to over a thousand or so. It may be almost stationary, or may travel up to 1000 miles in a day, having a duration of anything from 1 to 5 or more days.

In general, the weather may be expected to approach from the west to south-west, and the barometer to fall. The wind backs to south or south-west, bringing fine rain and wind. As the centre approaches, the temperature and humidity both increase, and the wind, veering south-west or west, gives persistent rain.

The passing of the depression is shown by the sudden movement of the wind and squalls to the north-west as the sky clears.

It should not be assumed that all depressions are copy book reproductions of this ideally theoretical example, as bad weather systems do not always follow this exact sequence, but the conditions described may be taken as representative of typical 'low' behaviour.

So far, we have been examining the behaviour of a depression in plan-view – that is to say, as if we were directly above and looking down on our schematic representation – but really, since there is a strong vertical development of cloud and movement of air, it should be a completely three-dimensional project, showing both horizontal and vertical occurrences as fronts approach and retreat across a given point of observation.

With the further aid of simple diagrams, we endeavour to separate the behaviour of the various fronts and to attribute to them individual characteristics which we can recognise as a typical depression forms and dies.

Fronts

A front, then, is an area at which weather conditions vary sharply and along which rain usually falls. It is the meeting place of different air masses with temperature differences which do not mix but continue to move parallel to each other, either in the same, or in opposite directions.

For example, warm south-west air may meet cold polar air, and a front is thus created along which the two types of air fluctuate in strength and oscillate in direction until the wave thus caused either levels out again or develops into the kink or bulge which deepens into the kind of depression recently discussed.

The development of frontal conditions is called *frontogenesis,* while the deterioriation of frontal conditions is called *frontolysis.*

By way of definition, we can say that when polar air overcomes the fluctuating frontal division it is called a *cold front,* and when the sub-tropical air advances it is called a *warm front.* Each front produces its own circulation of air within the extensive main body of the weather system, and on all fronts rain may be expected.

When a cold front overtakes a warm front and closes the depression, it is called a *warm occlusion.* When a warm front overtakes a cold one and closes the depression it is said to be a *cold occlusion.* In schematic diagram form, the occlusion is shown as a frontal line having alternate tooth-and-dome projections facing in the direction in which the front is travelling.

A *stationary front* is one in which there is practically no horizontal movement. In diagramatic form, this is represented by a line having both tooth and dome symbols facing alternate directions.

Warm front characteristics

When a mass of warm air advances on a mass of cold air, it will, by virtue of its lesser weight, rise over the cold air in a long gentle incline, causing a particular sequence of clouds which we are able to identify as they occur as ranging from high formations to low. They are responsible, therefore, for the belt of steady rain and a falling ceiling towards the centre of the depression.

By reference to previous weather charts, an observer can see that a depression is approaching his observation post, which at the present time may be immersed in a mass of cool air, and he may prepare himself for the sequence of events which is to follow (Figs 16a and 16b).

A mass of warm air, approaching from many miles away, advances on the cold air at present prevailing over the countryside, and since warm air will always rise, this oncoming mass is forced up above the prevailing cold air in a steep incline. At that point the warm front is established.

The barometer falls, indicating a steady fall in pressure, since the warm air travelling above it is less dense than the cold air hitherto affecting the reading.

The rising air forms a cloud sequence beginning with cirrus, and the direction from which the cirrus is spreading is the direction from which we may expect the warm front to approach.

In steady progression, the clouds thicken, and at about 26,000

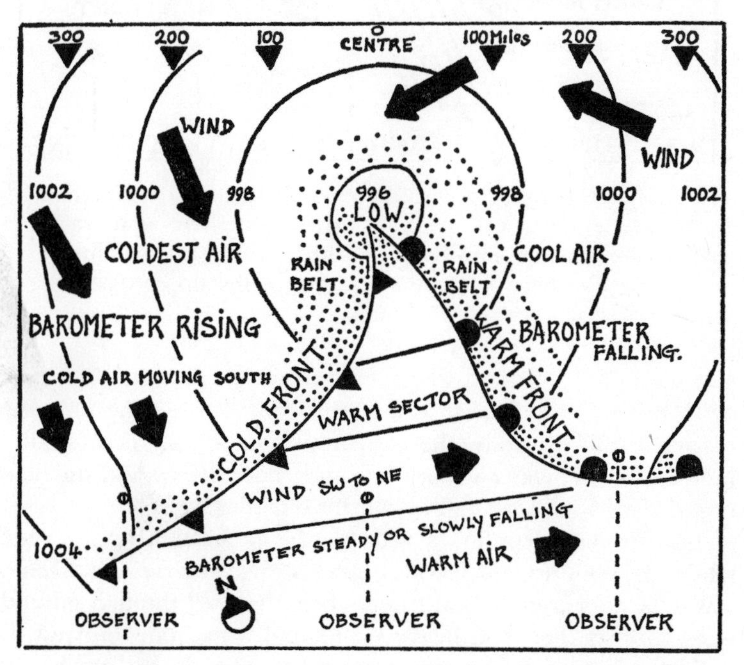

16a *Conditions in a theoretical depression.* The depression deepens and becomes very active, bringing winds increasing to gale force, and heavy rain. The diagram represents a plan view (looking down) of the conditions at present under discussion.

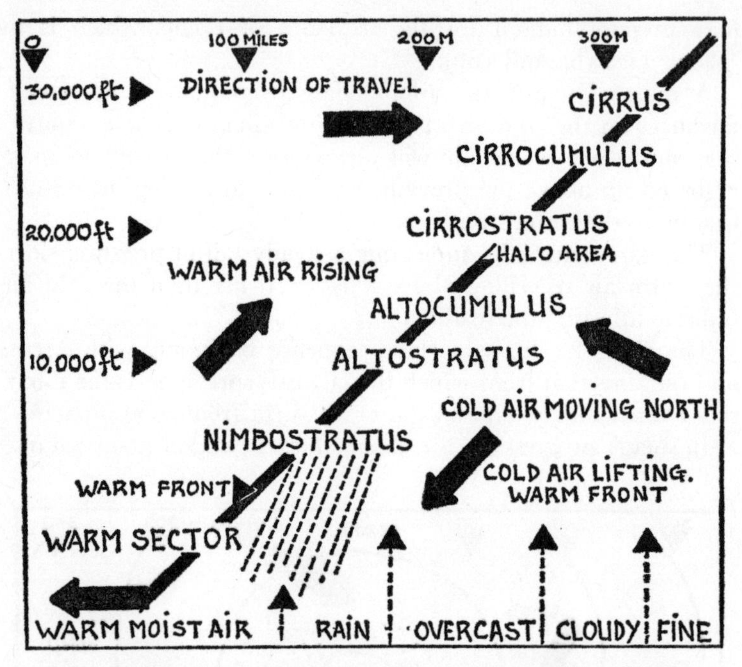

Fig. 16b *Conditions in the warm front region.* The diagram represents a vertical picture of the atmosphere as a warm front approaches, and may be assumed to be the conditions existing along the warm front region of the upper diagram.

feet merge into cirrostratus, a change which may be some 300 or so miles in advance of the warm front. But all the time the barometer falls steadily, the temperature rises, and the humidity increases as the water vapour becomes more dense; and the halo round the Moon that night confirms our diagnosis.

The sequence continues thus as the depression approaches, while at about 12,000 feet, altostratus gives the Sun, or Moon, a watery appearance, as if it were being viewed through ground glass, and as the dark, heavy clouds move in from the west, it indicates that rain is near at hand. Soon after the approach and retreat of the altostratus comes the grey, menacing nimbostratus,and rain begins to fall over the entire area occupied by the front.

Sometimes layers of stratocumulus may herald periods of

sunny weather as the warm front passes and the warm, moist air of the warm sector covers the area.

Warm sector characteristics

As the centre of the 'low' arrives, the barometer, thermometer and hygrometer remain steady or with small variations throughout the movement of the warm sector as it passes spreading rain.

During the winter and early spring months, the moist Atlantic air of the weather system giving rise to the warm sector will probably develop low stratus or ground fog, which may last well into the following day. Otherwise, there is a fairly stable outlook during the summer months, with small cloud formations and periods of sunny weather, while the general tendency is towards cumulus by day and clear skies by early evening.

The barometer at the warm sector, near the centre of the depression may be reading in the region of 980 millibars.

Following somewhere behind the warm sector is the cold front, an irregularly-shaped zone of frontal air derived from the cold air mass being drawn round the rear of the depression, and feeding another sequence of weather into the life-cycle of the system.

Cold front characteristics

By reference to our theoretical diagrams, it is possible to visualise the conditions which now probably exist as the whole weather system moves across the observer.

For example, there is a mass of cold air creeping around from the east side of the system to the west, and turning inwards towards the warm sector, which by now is tending to become narrower as the cold front created by the oncoming cold air advances towards the warm front at a slightly faster rate of travel.

The stratocumulus of the warm sector now gives way to cumulonimbus as the advancing cold air undercuts the warmer air in the warm sector and forces it upwards, forming dense, turbulent cloud. According to conditions, there may be hail, rain and thunder (Fig 17).

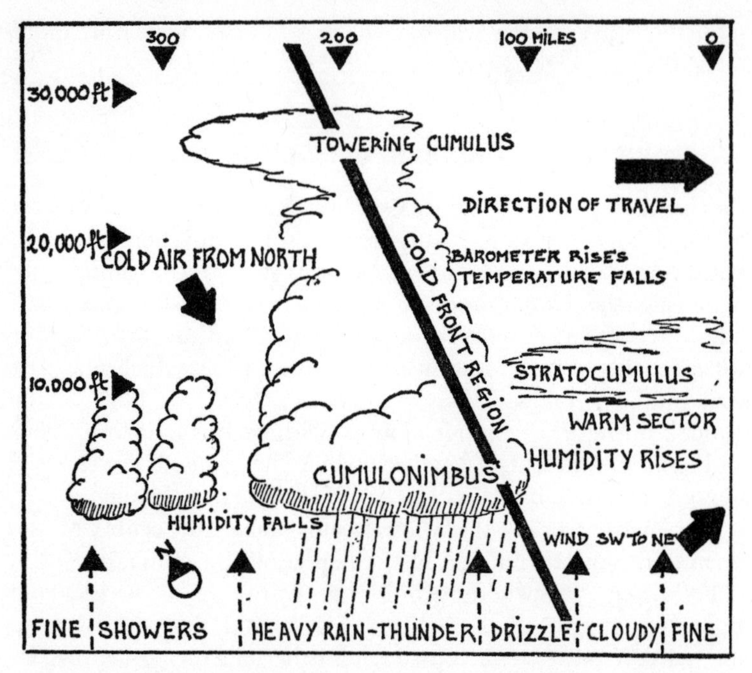

Fig. 17 *Conditions along an active cold front.* The diagram represents a vertical picture through a cold front where the lifting of the warm sector is taking place. This may be assumed to be the conditions existing along the cold front region depicted in the theoretical depression shown in Fig. 16.

At the approach of the cold front, the barometer begins to rise due to the greater density of the cold air; the temperature falls, and the humidity decreases as the nimbus passes.

Whereas the warm front is extensive, the cold front is comparatively narrow, extending hundreds of miles in length and only some 40 to 50 miles in width. As the front passes with its squally showers, the wind veers, perhaps from west to north-west, and the barometer rises. The temperature drops, although it may be only slightly, and the cold air brings a period of good visibility and sunshine.

When there is not much difference in temperature between the two zones of air, the warmer is lifted slowly and clouds thicken only slightly.

Light rain may fall but the showers soon pass to reveal blue sky.

When the depression is fast moving and the cold air behind is much colder than the warm sector, the lifting of the warm air is more violent. There is a short period of heavy rain, sometimes with thunder and lightning from the clouds which have grown rapidly into cumulonimbus.

From these indications we may conclude that the final stage of the depression has begun.

Occlusion characteristics

As the cold front progresses, it catches up with the fast travelling warm front and begins to overtake it, moving from the centre of the 'low' outwards. Diagramatically, we can illustrate the process by drawing a frontal line and marking it with alternate tooth-and-dome symbols to indicate the filling-in of the sector as its warm air is forced upwards from ground level and the following cold air pushes forward towards the cool air over which the original warm front was formed (Fig 18).

Although the depression is now declining, winds and rain are still to be expected, and it often takes several days for a depression to extinguish, by which time the next one in line has already formed somewhere, either out to sea or over land; for example, depressions over the north-west Europe area which penetrate into the Baltic or Scandinavia also include the British Isles in their distribution of storms and poor weather, or we can experience a whole succession of depressions coming in from the Atlantic.

As the occlusion continues, the rain caused by the cold front diminishes and the clouds break up and drift away. The occlusion thus completes its process of re-establishing the equilibrium of the amosphere.

At the approach of the occlusion, the barometer drops at first and then begins to rise again. Temperature remains steady, the humidity varies a little and the wind veers sharply as the occlusion passes over.

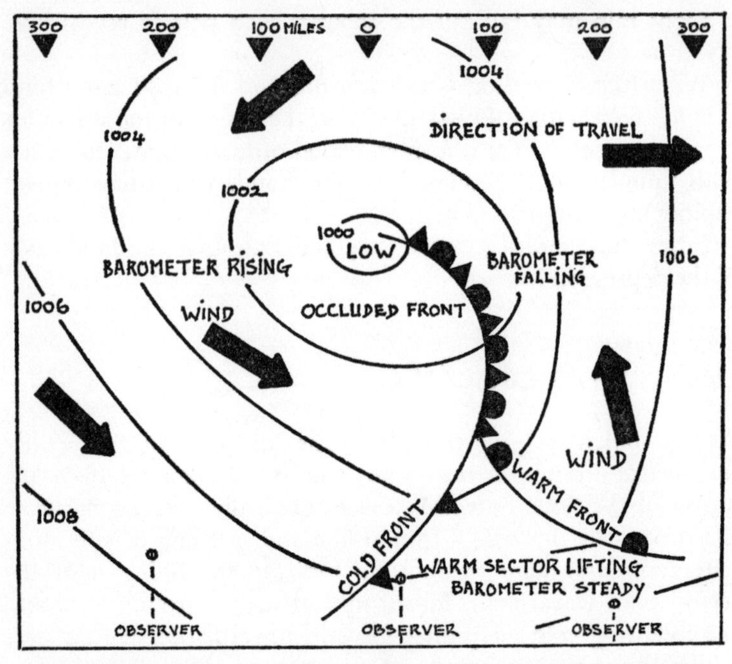

Fig. 18 *Conditions in an occlusion.* The depression is slowly filling in, but winds and rain are still to be expected. At the approach of the occlusion, the barometer falls at first and then rises with the passing of the front. Temperature remains steady. Humidity varies little. It may take several days for the depression to decline.

Warm and cold front occlusions

The process for the two types of occlusion is exactly as described except for the final stages. The cold air of the cold front pushes the warm air of the warm sector entirely off the ground as it meets and joins up with the cold air over which the warm front is rising. It can now take one of two courses open to it, depending upon the degree of coldness in relation to the air which it overtakes.

If the *advancing* cold air is *colder* than the air upon which it is advancing, it will undercut it and proceed to cause it to rise by virtue of the fact that the *overtaken* cold air is warmer by comparison than the *overtaking* cold air. This is the cold occlusion.

If, on the other hand, the advancing cold air is *less* cold than the air upon which it is advancing, it (the advancing cold air) will rise over it because it is warmer than the air which it has overtaken. This is the warm occlusion.

The weather which follows the passing of a depression will depend upon the behaviour and influence of surrounding weather systems, as may be revealed by the current weather map.

Experience of depression characteristics throughout depends entirely upon the position of the observer in relation to the centre of the system; for example, anyone on the west side of a depression finds that in general the winds, blowing anti-clockwise round the centre, produce an air stream from a northerly direction. This air is approaching from progressively warming seas as it moves southwards, and is therefore inclined to produce thermal action over the sea. An observer's experience of showers, clear skies, dew, frost or fog will therefore depend upon his position in the British Isles, and, of course, the influence of the season, since the showers occurring may be of snow and not of rain if it is winter.

If the observer is on the south side of a depression centre, he will experience the warm front, warm sector, cold front and occlusion to a degree depending upon his closeness or distance from the centre, and will find winds blowing south-south-west ahead of the warm front, west-south-west behind it and north-west behind the cold front.

Since most of our depressions cross the British Isles to the north after travelling upwards from the Atlantic, the south-westerly air reaches us having experienced the cooling effect of the ocean, and is not only cool to cold but is also moist, giving rise to stratus or fog by night and day in winter, spring, summer and autumn.

Our experience of any weather system is not a copy book pattern repeatable by a stereotype cloud sequence or predictable temperature readings. Weather is variable and at times unpredictable in its behaviour, and any set of weather maps will show that no two weather patterns are identical.

Fortunately, all of our weather is not dependent upon areas

of low pressure, and a good deal of variation is provided by the high-pressure systems which bring us an occasional ray of sunshine.

The anticyclone

The anticyclone is a wind system having an area of high pressure at its centre, with the wind running clockwise in the northern hemisphere and anti-clockwise in the southern hemisphere. Because of the high barometric reading of the anticyclone, the weather system is known as a 'high' in meteorological circles. It is generally expected to bring about settled to fine conditions, depending upon the season, the source-region of the break-away anticyclone (that is to say, whether it is Maritime or Continental). The topographical influence of low-lying areas are subject to frost and fog, high-level ground being prone to stratus and the creation of thunderstorms from the heat radiated by large land areas.

In a schematic diagram we can express an anticyclonic area by drawing a rough circle around the centre of high pressure and marking in the barometer reading – for example, 1034 millibars. Outside and around this circle we can draw others to represent the theoretical connection of all weather stations having the same pressure readings, and from this we can see how each circle reduces in value by about 2 millibars, so that the first circle outside the centre one reads 1032, the second circle reads 1030, the third 1028, and so on, until, perhaps, the pressure readings continue to drop as the circles progress outwards into other weather systems and we find them merging into the low-pressure lines of an adjacent depression. (Fig 19).

The velocity of an anticyclone is slow and may remain over one zone for several days, giving fine, clear weather in the summer with a little cloud near the centre although cloud and rain may occur at the edges of the system.

Normally, temperature decreases about 5°F for every 1000 feet of ascent, but in the winter anticyclone this rule does not stand fast. While the decrease is normal for several hundred feet, it begins to regain heat rapidly for some height after this and then decreases again, thereby creating a layer of warm air be-

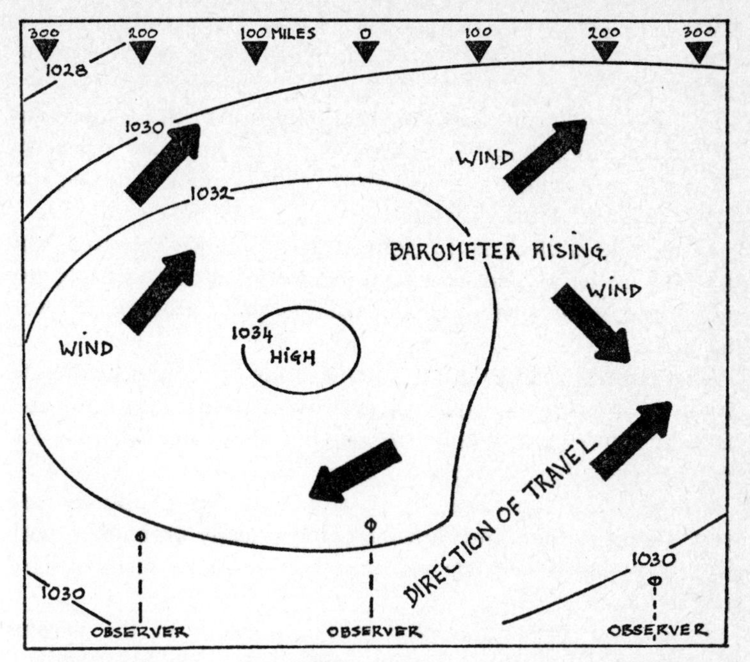

Fig. 19 *Summer anticyclone.* The exact opposite of the depression is the anti-cyclone, in which the wind direction is clockwise and the isobars are far apart with the highest pressure reading at the centre of the 'high'. In summer, the extension of the permanent high pressure area of the Azores brings in fine hot weather.

tween the colder air from the surface and the colder air of the outer atmosphere. This is called an *inversion*, and assists in making for stability, as any rising air unable to proceed through the inversion layer will sink back.

Should the rising air produce cloud, it will not ascend beyond the inversion level and will spread out, giving a low ceiling. As such items as smoke and dust are also unable to escape, poor visibility may be expected.

During the winter, the anticyclonic periods are created in the polar air travelling southwards, or originating from the powerful and persistent anticyclones centred over the Azores and spreading north-east.

During the summer, anticyclones arrive from the warm continents or are break-away high-pressure areas from the subtropical Azores anticyclone, all of which bring us fair to very good weather.

Autumn and winter anticyclones may bring stratus and rain, or fog due to the stable conditions of the air mass when the air near the ground is cooled below its dew point and becomes visible as cloud. Humidity increases.

If the sky is clear and bright, a frost may be expected, and it would be as well to watch the behaviour of the wet and dry bulb thermometers for indications of dew point around the freezing mark.

The Azores weather in the summer may bring us a heat wave which lasts for several days or even weeks, with uncomfortably high humidity and thunderstorms. But in winter we may experience extensive fog or stratus at low level.

Other air masses are dry Continental types, approaching from across Europe and showing a high steady barometer, with dry, fine weather heat waves in summer and long spells of hard frost in winter.

In general, the differences in the pressure of the various observation points within the anticyclonic area are gradual, and winds are light and variable with calm weather and little rain, but the differences between winter and summer anticyclones are marked by long, clear days and high temperatures in the latter, and short days and frost or fog in the former.

The anticyclone is often welcome, but more often in summer because of the stable weather it brings, but is frequently unwelcome, more often in winter for the periods of cold, unpleasant weather it often brings.

The anticylone, or 'high', tends to remain for some time once it has established itself over an area, and is at the best of times slow-moving and strongly opposed to change, so that it might experience the approach of a succession of depressions that all attempt to displace it but are obliged to sheer off in another direction without having caused much of an impression.

Detection of pressure systems

If we examine a schematic diagram which shows an area of low pressure with several circles of pressure readings around it, adjacent to an area of high pressure with several circles of pressure readings around it, we shall see that since the winds blow anti-clockwise in a depression, we can stand at any point throughout the system and, with our back to the wind, safely declare that the centre of low pressure is on our left and that any high pressure area is on the right.

Subsidiary systems

Developments of high and low pressure areas are irregularly spaced around the globe, working continuously within the parent air masses that produce them. And somewhere between these major atmospheric disturbances there are smaller developments which form troughs, wedges, cols and secondary depressions, any of which may either decline and die out, or develop into an area of major importance in its own right (Fig 20).

The lines which we have already used on our schematic diagrams to connect weather stations having the same barometric pressures are called *isobars,* and the diagrams they form when properly drawn into place on a weather map are called *isobaric patterns.* If we now draw a theoretical diagram, or isobaric pattern of pressure lines as they might appear in relation to 'high' and 'low' areas, we shall clearly make out the following detail:

A col

A col is a region occurring between two anticyclones and two depressions arranged alternately, and arises as the result of the clockwise motion of the two adjacent anticyclones which form between them this area of light air originating from different directions, so that it becomes an area of relatively low pressure between them.

It may be likened to a valley between two hills, giving rise in summer to conditions promoting thunderstorms, and favourable for the formation of fogs in winter.

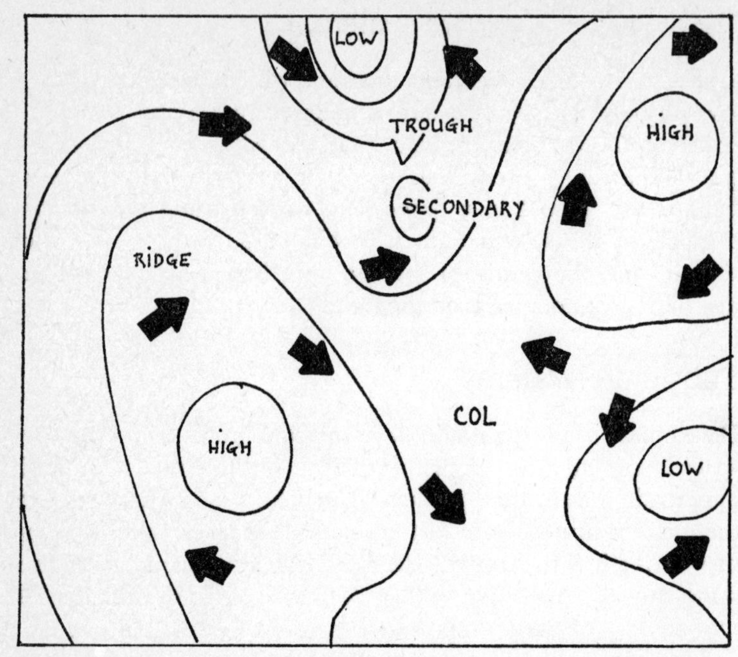

Fig. 20 *Subsidiary systems.* Somewhere between the major atmospheric disturbances there may be smaller developments which form troughs, ridges, cols and secondary depressions. The diagram is a theoretical representation of these systems.

A trough

A trough of low pressure is indicated by an extension of the isobars of a low pressure system and forms a sharp V. The apex of this V marks the region of abrupt changes of wind direction, and usually lies between two areas of high pressure.

A secondary depression

A secondary depression forms as part of the main depression system, and may by development become greater than the parent system.

It is a fact that the isobars distinguishing a depression do not follow symmetrical circles, but may form into bulges where the wind in that part of the system deviates slightly from the general circulation. This bulge may form its own area of low pressure, and when that happens it is said to be a secondary depression.

In winter, it may give rise to snow, rain or gales. In summer, when the secondary is comparatively shallow, thunderstorms may be expected.

A ridge or wedge

A wedge, or ridge, of high pressure on the map is an extension of isobars from an area of high pressure. The formation is V-shaped, although not as sharp at the tip as the V in a trough.

The wedge usually extends from the edge of the anticyclone and between two areas of low pressure; the conclusion to be drawn is that the wedge is a dry interval sandwiched between two rainy spells.

The wedge is preceded by a region of fine weather with clear visibility and north-east winds. When the winds begin to back, we may conclude that a new depression is on the way and that rain will shortly fall.

The wedge itself may develop into an anticyclone to continue the existing conditions or to act according to seasonal characteristics.

During the winter, the air mass causes widespread cumulus, particularly on the east coast, while inland there is stratus and fog.

Coming in on summer air, the wedge produces east coast fog and stratus, and small cumulus inland.

The eternal problems

From our studies so far, we have seen how the balance of the atmosphere is affected by the heating effect of the sun, and how different kinds of air bring different experience of temperature, pressure and humidity.

We have seen that in the aerial ocean in which we live, the

gases comprising the atmosphere are each as important as one another, and that the most variable and useful part of our atmosphere is water vapour, since life in all its forms requires moisture in some degree, however small or however large. The continuous chain of events such as evaporation, cooling, condensation and precipitation produce our experience of clouds and rain, and ensure our everlasting supply of the water we so consistently need.

It has become evident from our experience of the origins and circulation characteristics of our weather systems that while the detection of an air mass on the move is not a great difficulty, anticipation of its probable movements is by no means as simple, and in some cases jolly nearly impossible; and since the weather is caused by a number of factors, all of which are fluid in their behaviour, it is small wonder that forecasting is often a hazardous task. Even the basically simple exchange of air from cold to hot areas of the globe in vertical circulation is interfered with by the globe itself being on the move and distressing the path of straight exchange, so that the winds spiral rather than flow directly from one zone to another. This is, of course, a known factor, and can be charted approximately and explained in detail, but even the basic world pattern of winds is upset by the existence of oceans and continents which produce their own influences on the flow of air by virtue of contour and friction, and by their temperature. Again, if all the globe were composed of water, all air would be of the Maritime variety, or if it were composed entirely of land, all air would be of the Continental variety; as it is, we are given a selection of Maritime and Continental air, blowing hot or cold according to direction and season.

Not only is a theoretically perfect flow of air not possible for the aforementioned reasons, but it is also affected by strong temporary systems of high or low pressure, all battling for an existence and for the most scientifically advantageous position on the globe in which to feature their life cycle, so that they are always in various stages of creation and deterioriation, and wander about the Earth's surface giving us our extreme changes of weather.

Between the polar regions and the tropics, weather conditions are variable from day to day, and sometimes from hour to hour, but in the deserts, the tropical rain forests and the polar wastes, the weather is constant and therefore predictable, since any sudden change in barometric pressure, temperature or humidity is a reliable indication of imminent unusual conditions.

The prevailing winds of the British Isles mostly come in from west to east, so that fine weather systems and bad weather systems reach us by way of the Atlantic and our experience of winds so associated is concerned with variation of strength and direction, of which direction and speed of depressions arriving from the Atlantic prove to be the most difficult to predict since they are apt to change course without warning. Alternatively, a depression may quickly fill in and pass by with nothing more to indicate its passing than a slight breeze, or it may intensify and sink us in a zone of foul weather measuring several hundreds of square miles in all directions.

On the other hand, the anticyclone is not always the centre of fine weather it is often made out to be, and in the winter we may find ourselves transferred from the rain-dull-bright-dull attitude of a sequence of fairly friendly depressions, into the bright frosty weather of a none-too-friendly anticyclone, which may turn even less friendly and sink us in widespread fog or low-level stratocumulus for days on end.

The summer anticyclone is a much better proposition and it is to this that we look for our most glorious weather, notwithstanding the fact that thunderstorms often follow long periods of fine hot weather.

Our little island, between the Atlantic Ocean on the one side and the vast European land mass on the other, surrounded by water, its land area distressed by mountain ranges, hills and valleys, and being just outside the belt of settled conditions enjoyed by the south Atlantic and the Continent, provides the weather forecaster with a complexity of conditions, possibilities and probabilities without parallel.

As an indication to our day-to-day weather, most of us try to depend on a swift glance at the sky to see whether it is clear or cloudy and, if it is cloudy, to determine if the clouds are rain or fair weather clouds in layers or in heaps, high or low. Then we test the 'feel' of the weather by its impression of damp, cold or warmth on our face and hands. If there is a breeze or a wind, we may look about us to find out from which direction it is blowing, and from this decision determine, in our personal experience, if we usually get rain from that quarter or the good chance of a fine day. By these simple observations, many people have become remarkably wise in forecasting their local weather for the day, or even the following day, without ever knowing the name of a cloud, the pressure of the air or its relative humidity. Their experiences and observations over a long period in the area have provided them with all the tools they need to decide on a trip to the coast or whether to take an umbrella on a local shopping expedition. The uncanny weather-wisdom usually attributed to seafarers and shepherds is a skill of this kind.

But for the reasons already mentioned, and probably a good many more besides, the professional meteorologist has a considerably more difficult task in relating his skill and the details of hundreds of weather reports to a chart of the entire northern hemisphere, and from it producing a forecast of conditions likely to develop during the next 24 hours.

In the following chapter, we shall be able to bring all our recently acquired theory and practice to bear in our efforts to chart the weather for ourselves.

7/CHARTING THE WEATHER

'If the matting on the floor is shrinking, dry weather may be expected. When the matting expands, expect wet weather.'

Since the state of the atmosphere may change from hour to hour, it is necessary for the weather forecaster to be in possession of hundreds of weather facts at any one moment and to be able to revise them as necessary when fresh information indicates new developments in prevailing conditions, so that he may weigh up the probabilities of future conditions and accordingly prepare his forecast.

While the barometer provides a fairly accurate pointer to the probable local conditions within the next few hours, a widespread forecast can only be made with the aid of a large collection of facts and figures assembled from other areas of the country, and, indeed, other areas of the world. As we have seen, the weather is an all-over occurrence, and the originating source affects a country many hundreds of miles away by the characteristics it imparts to the air mass which leaves it to venture across the globe.

Information is gathered from observing stations situated on land in permanent buildings and huts, from the Weather Ships at sea, from aircraft making periodic and altitude flights, from radio satellites, from radio sonde balloons, and from hundreds of private individuals all over the world who operate home-based weather observation stations as a serious hobby.

When these reports are received at the Central Forecasting Office, the information from them is transferred to weather charts. These are maps covering the principal forecasting areas, showing no place names or topographical detail, and usually covering the British Isles, north-west Europe, the Azores, Iceland and south Greenland.

From the many stations concerned, the charts show reports in a kind of meteorological shorthand which surround each station's position on the chart. The code expresses pressure, temperature, cloud, wind direction and speed, humidity, visibility and a knowledge of the upper air, to mention a few of the basic items required. Before a weather forecast is made, it is necessary to have all the current information about the condition of the weather in all parts of the northern hemisphere (since this is the area with which we are concerned) and to have it accurately plotted on the chart for a given time of day. A chart providing information concerning the overall weather conditions at any one time is known as a 'Synoptic' chart.

The centres of high and low pressure, the frontal systems and the lines of equal pressure (isobars) are drawn in after the station reports have been entered, and then the forecaster prepares what is called a Prebaratic Chart, which is his opinion of the weather situation at the end of the forecast period. Even then, it is necessary to examine frequent reports and to analyse them for subsequent changes in the weather situation so that fresh developments can be plotted on new charts.

Even with the comparatively recent use of computers and satellite transmissions, there is still no such facility as an instantaneous weather report in which all the factors from all the stations can be plotted at once and a forecast produced simultaneously, so that any weather map and any weather report or forecast that reaches us via newspapers or television is already out of date, and is most probably being reshaped with fresh information at the very moment of our seeing it. And so meteorology remains an inexact science with forecasts rather than predictions, probabilities rather than certainties and educated guesses rather than resolvable statistics, all of which allow ample scope for the amateur weather-watcher to work on his own initiative while the professional meteorologists and the scientists pit their technical skills against the restless air in which we live.

Weather information

From all the sources of information with which the professional meteorologist is in contact, details of the weather prevailing, certain diminishing symptoms and certain future indications are all collated into one overall picture, so that a forecast may be prepared. This information arrives from radio sondes, weather ships, satellites, computers, special met flights, private observers, and world-wide weather information services which maintain a 24-hour watch throughout the year. Radio transmissions provide synoptic weather messages on an international scale, using an agreed code which can be interpreted easily by the receiving station into the notation used to express all the elements indicated.

A synoptic weather message is composed of 5-figure code groups which, when analysed, reveal the details that are then entered on the synoptic (or weather) chart. The completed report includes information about horizontal visibility, barometric pressure, temperature, dew point, wind speed and direction, precipitation, sunshine, humidity, cloud type and amount, the past weather and barometer tendencies. In addition to station reports of this kind, it is necessary to accumulate similar information about the upper air, and this, as we have previously seen, is provided by radio sonde observations tracked by radar stations.

It is the combined effect of these elements at any one time that is the state of the *weather,* while the combined effect of their average behaviour over a long period on any one place is the *climate* of that area or country.

An example of coded weather reports is to be found in the Daily Weather Reports of the British Meteorological Office on the reverse of the weather chart, and although a section of this is reproduced here for reference, it would be of value to future studies to obtain a few Weather Reports from H.M. Stationery Office (Fig 21).

We do not need to interpret the 5-figure groups printed thereon, but to give a general idea of the lay-out of the report the following guide should be studied.

The Code number FM 11.D in the top left-hand space indicates that the information is derived from Land Stations, while the Code number FM 21.D in the lower left-hand space indicates that the information is derived from Ship Stations.

Each reporting station has its own identification number, and this appears against the station name under the column headed iii. Ships are identified by their latitude and longitude, which appear under the two columns 99 LaLaLa and QcLoLoLoLo

DAILY WEATHER REPORT OF THE BR

No. 40963

Code FM 11.D — **OBSERVATIONS at 12h.** 15 JANUARY 1974 — **OBS**

Station	iii	Nddff	VVwwW	PPPTT	$N_hC_LhC_MC_H$	$T_dT_d$app	$8N_sCh_sh_s$	$8N_sCh_sh_s$	$8N_sCh_sh_s$	iii	Nddff	VVwwW
Boscombe Down	746	72821	82806	12212	784//	09319	85711	83814	86628	746	72917	80028
Hurn	862	72520	62016	13313	5843/	11114	83818	83640	85360	862	52808	70012
Thorney Island	871	82423	62025	13612	853//	11218	84708	86710	88618	871	62809	80012
Manston	797	82320	74506	12312	873//	11209	85707	88709		797	72612	84025
Gatwick	776	82424	70586	12012	873//	11008	81709	87711		776	72614	81028
Kew	775	72313	75036	10814	764//	10209	87716	86645		775	62611	70036
Heathrow	772	72516	70806	10913	785//	11211	83822	86628		772	62710	80028
Honington	586	82517	33622	07513	854//	10003	83618	88625		586	72718	82038
Wattisham	590	72220	50626	08612	754//	11109	86612	87624		590	72615	82016
Gorleston	497	72218	60022	06613	755//	10206	87620			497	42718	66021
Aberporth	502	52526	69036	11509	32432	05144	83815			502	52716	74022
Ross-on-Wye	627	52519	82025	10112	38538	07235	82825			627	52408	77021
Filton	628	32524	70016	12611	38500	07236	81820	83656		628	52611	70011
Rhoose	715	22524	66016	12910	11471	07344	81815			715	52613	66016
Chivenor	707	62618	80016	15110	21432	08242	82815	85359		707	52913	75018
Mount Batten	827	72813	75025	17211	6544/	08248	86618	87360		827	72510	66022
St. Mawgan	817	72720	62026	16510	5537/	08144	81707	85645	87359	817	73013	70028
Scilly	804	72518	62026	18211	7567/	09248	87645			804	82704	62026
Culdrose	809	72815	72026	17310	5543/	08245	81712	84622	85650	809	83005	69025
Guernsey	894	82424	56216	17711	872//	11219	84704	87706		894	72712	68026

Code FM 21.D — **OCEAN WEATHER STATION REPORTS at 12h.**

O.W.S.	$99L_aL_aL_a$	$Q_cL_oL_oL_oL_o$	YYGGIw	Nddff	VVwwW	PPPTT	$N_hC_LhC_MC_H$	D_sv_sapp	$8N_sCh_sh_s$	$8N_sCh_sh_s$	$OT_sT_sT_dT_d$	$IT_wT_wT_wtT$	$3P_wP_wH_wH_w$	$d_wd_wP_wH_wH_w$
A														
B	99570	70516	15124	72410	97858	93958	78400	00713	82812	86625	0//62	10249	30403	30906
C														
D														
E														
I	99589	70190	15124	52727	97157	93404	23462	61240	81915	83070	0//52	10872	30708	25011
J	99526	70198	15124	62614	98258	10006	29462	00122	82915	84070	0//02	10990	31010	29908
K	99450	70160	15124	81211	97585	20710	854//	00826	84713	88640	0//09	11247	30501	23005
M	99658	10018	15124	71305	97258	76305	5947/	00208	85918	/////	0//02	10704	30000	99505

Fig. 21 *Section of the Daily Weather Report*

respectively. Other than these indicators, we have the following for Land Stations:

N dd ff = total cloud, wind direction and speed

VV ww W = visibility and weather

PPP TT = barometer and temperature

NhCLhCmCh = amount of lowest cloud, type of low cloud and its height, type of medium and high cloud

TdTd app = dew point, barometer tendency

8NsCsh hs = extra groups to report amounts, types and heights of significant cloud.

Ship reports incorporate some of the above and the following additional information:

YY GG i=date, hour, how wind is measured

Ds Vs app = direction and speed of ship. Barometer tendency.
OTsTsTdTd = sea temperature, dew point
I Tw Tw Tw tT = sea surface temperature
3 Pw Pw Hw Hw = wave periods in seconds, wave height.

It can therefore be appreciated that, given the key to interpreting the figure groups, a complete synopsis can be obtained from the information given by them.

As we can see by reference to a small section of the weather chart on the reverse of the Daily Weather Report, each of the principal stations is accorded a position on the map by means of a circle, and this circle, although of minute proportion by comparison with its surroundings, is of the greatest importance to the forecaster, since it is from these small beginnings that the broad outline of the weather map is produced (Fig 24).

The Station Circle

The geographical location of a weather station is recorded on the forecast map by means of a circle. In and around the circle is entered all the weather information concerning that particular station, and for speed of handling, this information is transposed into a code notation consisting of letters, figures and symbols which indicate such information as:

i type of high cloud
ii type of medium cloud
iii barometric pressure at mean sea level
iv barometric tendency during past 3 hours (rising or falling)
v weather during past 6 hours
vi type of low cloud
vii amount and height of base of low cloud
viii dew point
ix visibility
x weather at time of observation (rain, drizzle, etc.)
xi air temperature in Centigrade (dry bulb reading)
xii (within station circle) total cloud amount
xiii (at appropriate position around the circle) wind force and direction

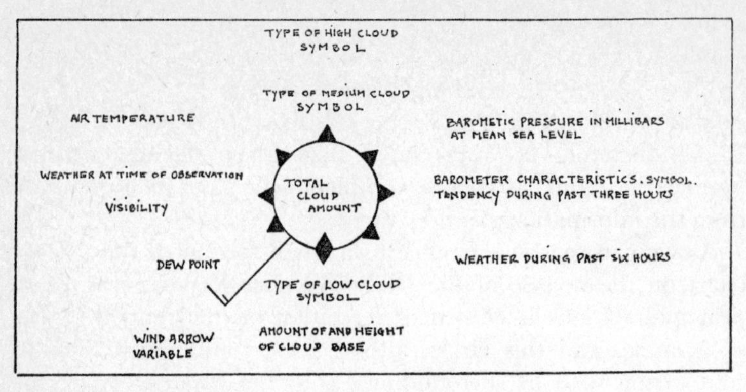

Fig. 22 *The station circle.* Each element of information has its own particular position relative to the circle

xiiii a dead calm is indicated by an outer circle around the the station circle.

As the example of plotting will reveal (Fig 22), each element of information has its own particular position in relation to the circle; for example, the temperature is indicated in a top left-hand position outside the circle, the barometric tendency in a right-hand position, cloud amount is *within* the circle, and so on. Since the wind is variable in direction, the wind arrow can be placed anywhere according to which direction it is blowing.

It becomes apparent that, given the information so far discussed in this book plus a selection of simple weather recording instruments and a collection of reporting code, anyone can, with a little practice, produce an acceptable do-it-yourself station circle in which all the elements are entered in their proper places and accurately represent all the conditions existing in the area of observation.

For the purpose of experiment, bring forward a copy of the Temperature Conversions from Chapter II, the Visibility Code, Humidity and Dew Point tables from Chapter III, the Sky Cover Code from Chapter IV, and the Wind Force Chart from Chapter V, bearing in mind also that we may need the North and South Cone signs to denote storms from those regions.

With the sample plot as a guide, it is possible to make several

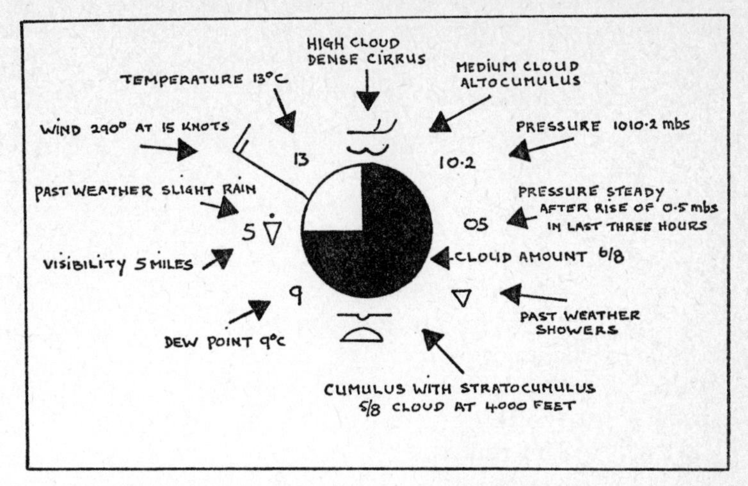

Fig. 23 *Practice plot.* This diagram shows exactly how each element is plotted.

attempts at constructing a station circle, varying the sky cover, wind arrows, visibility, pressure and temperature for each (Fig 23).

Remember that, whereas the actual wind vane indicates the direction from which the wind is blowing by heading its arrow point into it, the wind arrow on the weather chart turns its tail feathers towards the oncoming wind and its point towards the station. For good measure, one can occasionally add a storm cone for present weather.

As we shall see, there are a good many code letters and symbols used in practice. They are so numerous that it is not possible to record them all on the Daily Weather Report as sold to the public, so that the edition we see has been mercifully simplified for our convenience and understanding.

Full details of symbols are given in Meteorological Office Publication M.O. 515 *Instructions for the Preparation of Weather Maps,* and a few of the more common symbols and letters are now shown for immediate use.

More Weather codes and symbols

The Beaufort Notation

b	blue sky
c	cloudy
d	drizzle
f	fog
h	hail
i	intermittent (prefix)
j	in sight but not at station (prefix)
l	lightning
m	mist
o	overcast
o	slight (suffix)
p	showers
s	snow
z	haze
bc	partly cloudy
fs	low fog over sea (coast stations only)
ido	intermittent slight drizzle (prefix i = intermittent)
jr	rain in sight (prefix j = in sight)
kq	squall
pr	rain shower
prs	sleet shower
ro	slight rain (suffix o = slight)
rs	sleet
fg	low fog over land (inland stations only)
h (r)	rain and hail
ks	storms of drifting snow
k/s	heavy storms of drifting snow (generally low)
k/so	slight storm of drifting snow (generally low)
So/k	slight storm of drifting snow (generally high)
S/k	heavy storm of drifting snow (generally high)

Capital and small letters

Capital letters indicate that the phenomenon is heavy, e.g. R = heavy rain. Small letters indicate that the phenomenon is moderate, e.g. r = moderate rain.

The small suffix o indicates that the phenomenon is light, e.g. ro = light rain.

Repeated letters

Letters repeated mean that the condition is continuing, e.g. RR = continued heavy rain; rr = continued moderate rain.

Intensity

intermittent light rain = iro
intermittent moderate rain = ir
intermittent heavy rain = iR
continuous light rain = roro
continuous moderate rain rr
continuous heavy rain = RR

Cloud identification

Type	*Abbreviation*	*Code*	*Symbol*
cirrus	Ci	$C_H 4$	
cirro stratus	Ci St	$C_H 7$	
cirro cumulus	Ci Cu	$C_H 9$	
alto stratus	A st	C_m I	
alto cumulus	A Cu	$C_m 3$	
fine weather cumulus	Cu	C_L I	
towering cumulus	Cu	$C_L 2$	
cumulo nimbus	CuNb	$C_L 3$	
nimbo stratus	NbSt	$C_L 6$	
strato cumulus	StCu	$C_L 9$	
cumulo nimbus with anvil		$C_L 9$	

Pressure changes

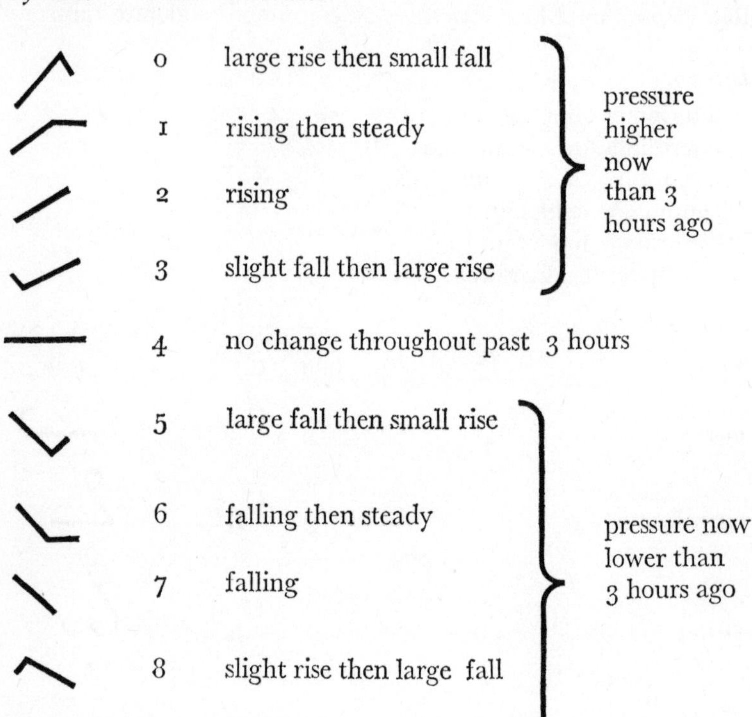

Symbol	*Code*	*Character*	
	0	large rise then small fall	pressure higher now than 3 hours ago
	1	rising then steady	
	2	rising	
	3	slight fall then large rise	
	4	no change throughout past 3 hours	
	5	large fall then small rise	pressure now lower than 3 hours ago
	6	falling then steady	
	7	falling	
	8	slight rise then large fall	
	9	not used	

Showers

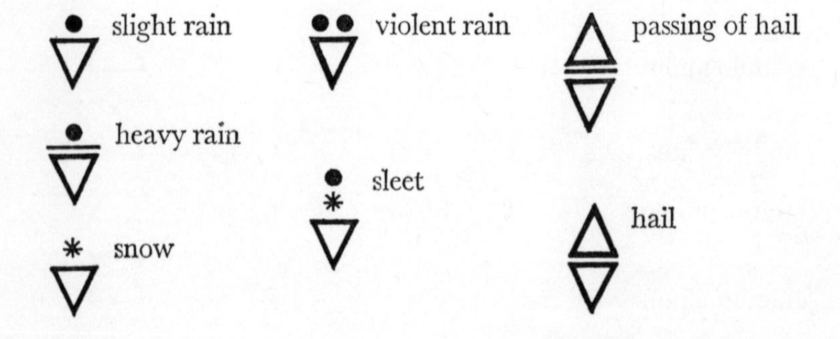

Miscellaneous

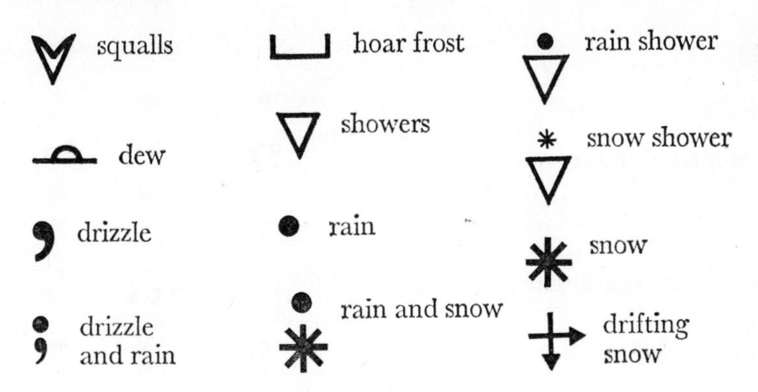

Past weather

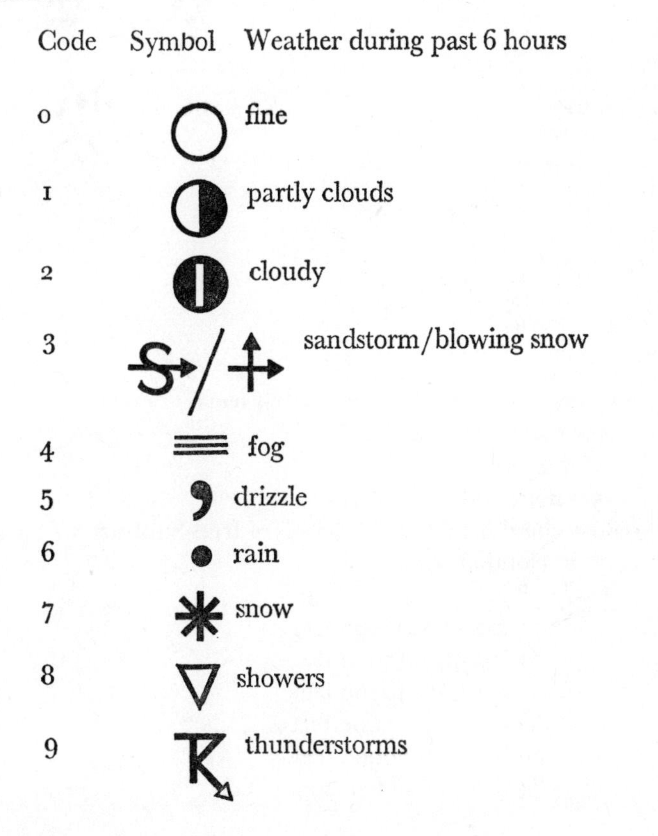

Present weather

	drizzle	*rain*	*snow*
light intermittent	,	•	*
light continuous	,,	••	**
moderate intermittent	, ,	• •	* *
moderate continuous	, ,,	• ••	* **
heavy intermittent	, , ,	• • •	* * *
heavy continuous	, ,, ,	• •• •	* ** *

precipitation within sight, distant from station)•(

precipitation within sight, near but not at station (•)

Height of cloud

Code
00 to 50=cloud height in hundreds of feet
e.g. 02=200 feet
25=2500 feet
51 to 55=not used
56 to 80=cloud height in thousands of feet. Subtract 50 to obtain cloud height.
e.g. Code 58 less 50=8000 feet
79 less 50=29000 feet
81 less 50=35000 feet
82 less 50=40000 feet
83 less 50=45000 feet
88 less 50=70000 feet
89 =above 70000 feet

Fog information

Mist ═ Haze ∞

Fog in patches

	thinning	no change	thickening
Fog with sky discernible			
Thick fog			

Cloud amount

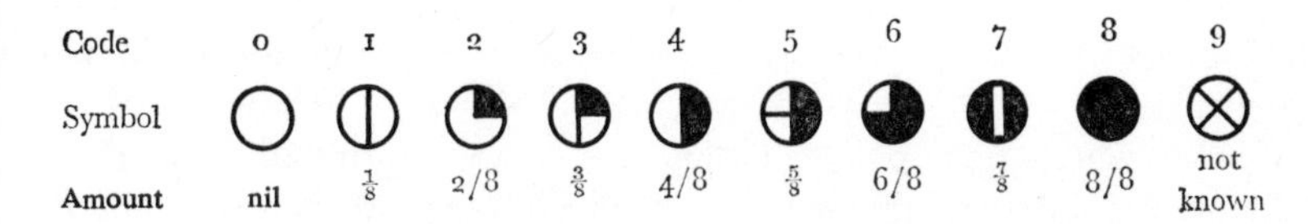

Thunderstorms and gales

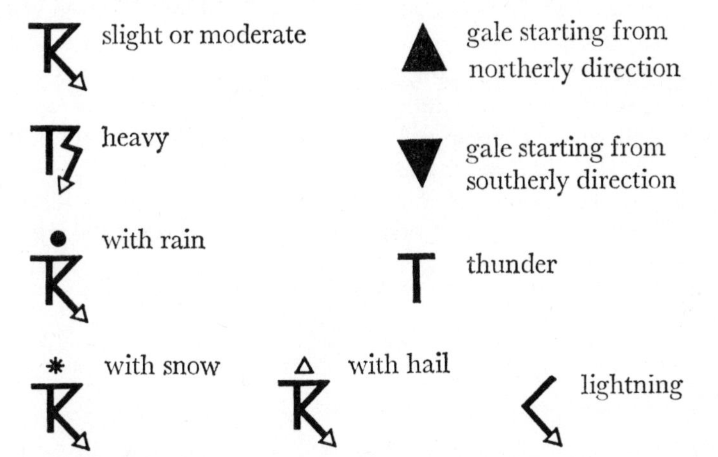

These indicators will allow for an improvement in plotting, since we are now in a position to make an entry for all the elements of information we require around our station circle.

It is now possible to visualise that if all the weather stations which contribute to a weather report are extended on to a large-scale map with all their information around them, an experienced meteorologist can read the overall conditions in much the same way as a chef can read a recipe book, or a chemist can interpret a prescription. So, then, the key to the forecaster's information lies in the careful preparation of each station circle, and, as we have said, with the information so far studied it is clear that everyone can, with practice, produce a reasonably accurate station circle for his own use.

Selective use of station information

The selective use of station circles for making analyses is of immense value to the climatologist as well as to the meteorologist, and comparisons are made by joining, by means of a line or lines, those places having similar existing conditions. For example, lines joining places of equal temperature are called *isotherms,* while *isonephs* join points of equal cloud amount. *Isohets* are lines joining points of equal rainfall over a given period, and *isohels* join points of equal sunshine duration over a given period.

For everyday forecasting we are principally concerned with *isobars,* for these join points (stations) of equal barometric pressure, and are the reference points of every forecast map for giving us the picture of areas of low and high pressure.

The effect arising from the production of isobars is similar to that which came about when we were discussing the theory of depressions and anticyclones in Chapter VI. Let us now find out how this is all applied to the official weather map.

Weather Maps

In the beginning, synoptic messages provide the professional forecaster with the detailed information from which the weather map is constructed. The result is something like a meteorological ordnance survey, using special symbols and notation to denote rain, depressions, anticyclones, pressure, temperature, humidity etc., instead of the conventional signs for railways, churches and roads.

Whereas the topographical map traces the contours of the

land, the weather map, by joining places of equal barometric pressure, distinguishes areas of high or low pressure over the entire hemisphere and provides us with a somewhat mobile pattern of the contours of the weather.

The meteorologist collates all the information contained in each station circle, studies all the elements concerned, and according to his understanding of the various situations, draws in the isobars to represent the conditions prevailing over certain areas. From this map he is able to construct the possible changes and path of such changes in the future, since by studying the map he is able to see the position of a depression or an anticyclone, the direction and speed of travel, and the general conditions within the system.

As the pressure systems are in a state of continual variation, no two maps are exactly alike.

If we now bring forward all the weather systems which we constructed in Chapter VI, we have ready-made isobars and weather systems which can be transferred to any area on a map of the northern hemisphere.

We can reproduce an entire depression near the British Isles, complete with frontal lines and imaginary pressure readings. Somewhere over the south Atlantic we can reproduce an anticyclone with its higher pressure readings; and with a little care we can draw in the conditions giving rise to a col, a trough of low pressure, a secondary depression and a ridge of high pressure, so that all the systems neatly interlock and provide us with our own practice weather map.

It seems now to be a simple matter to draw in a number of imaginary station circles somewhere along each line, and to indicate an appropriate wind direction (clockwise in the anticyclone and anti-clockwise in the depression), finishing off with an imaginary cloud amount within each circle.

Compare this result with the professional Daily Weather Report obtained from the Meteorological Office, and see how closely they are related (Fig 24).

As we can see, the overall result of drawing the pressure lines on a map produces a series of irregularly-shaped circles and almost parallel lines, intersected by the station circles surrounded by their individual observations noted in the abbreviations and code.

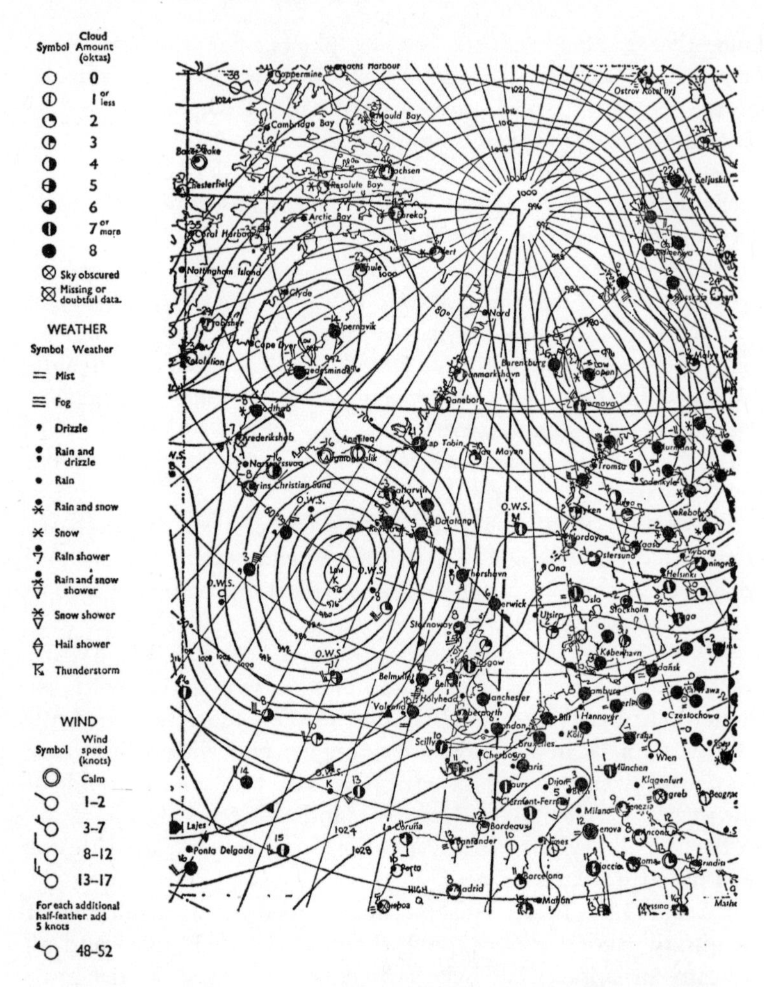

Fig. 24 *A section of the Daily Weather Report.* This weather picture shows a low pressure centre of 972 mb. in the lower left-hand side, and above that a secondary depression of 988 mb. To the right is a low pressure centre of 976 mb. and between the two, pressing in from the high pressure region below, is a ridge of high pressure extending from the 1008 mb. isobar. The 972 mb. 'low' shows the commencement of the occlusion at the centre, and extensive cold and warm fronts. Ready reference to the weather symbols which accompany the chart will provide all other information in detail.

Since the Daily Report is a much simplified version of the original, we are spared the task which confronts the meteorologist ever day of his working life, and we end up with an account of the bare essentials such as:

i sky cover;
ii wind direction and force;
iii temperature in C°;
iv pressure in millibars neatly placed on each isobar;
v symbols denoting present weather conditions at each station.

Naturally, we cannot miss out the all-important sea areas, since so much of our weather approaches from the Atlantic and so much of the character of moving air masses is influenced by maritime sources of the world, so it is necessary to include sea areas in the weather map.

The station circles which appear in the sea areas of the official Daily Weather Map are marked OWS A or B or C etc., and are, in fact, our weather ships which, like many land stations, transmit their coded reports by radio, and which are a vital part of the great observing network.

It will be seen that some isobar systems contain circles whose lines appear close together, indicating a rapid decrease of pressure and, consequently, of strong winds. This is the 'low' or depression, and it will be seen by examining the wind arrows that the winds are circulating in an approximately anticlockwise direction throughout the system.

The innermost bars may be marked at 980 millibars, increasing with distance at 2-millibar or 4-millibar intervals to something like 1008 millibars at the extremities.

The area of high pressure is shown by the isobars being arranged in similar manner in circles about each other, but in this case being much further apart, indicating a gradual pressure change and less strong winds.

Examination of the arrows shows that the wind direction is clockwise throughout the system, which the barometric pressure at the centre is, perhaps, 1032, falling with distance in 2-millibar or 4-millibar intervals, to something like 1016 at the extremities.

It will be seen how secondary systems, such as a depression, can develop out of the pressure system set up by the original one, and that several depressions, or anticyclones, can exist within one pressure structure.

Isobars are not allowed to cross but must form a continuous and complete route within their own systems. The skill of the meteorologist lies in knowing exactly how to complete these circles rather than allowing his isobars to wander off into some other area or to end up giving us no relevant information.

Since we have discussed the formation and appearance of all the principal weather systems, it remains only to remark that there are occasions when the isobars are seen to run in almost parallel lines across the map. These are called *straight isobars.* When this occurs the problem arises as to how the weather conditions can be determined from these 'non-system' arrangements, for there are neither highs nor lows nor troughs nor ridges to provide us with a mental picture of typical conditions and anticipated behaviour.

In this case, we have to determine the probable weather conditions according to the orientation of the isobars on the map. Should the bars lean towards the low latitudes from high latitudes *with* the wind stream they represent, one may expect rather cold weather, good visibility and showers. That is for lines running north and south in the northern hemisphere.

Should the lines be running in the reverse directions, one must expect poor visibility and mild weather.

The maps prepared by the Meteorological Office show the state of the weather at a particular hour, and are accompanied by a written forecast of the weather expected to develop from the stated systems.

Newspapers carry a forecast map which shows the expected weather conditions at a given time, and these are accompanied by other relevant information.

Although we never see them in general circulation, charts of the upper air are prepared from the balloon flights and upper air observations of which we have spoken in earlier chapters, and these too become part of the professional forecast. In upper air charts, the contours represent the height in metres above mean sea level at which the atmospheric level is 500 millibars. This means that isobars can appear having readings of 5400, 5460, 5520, etc., each reading representing the height at which the level is 500 millibars.

Colour indications on weather maps

The Daily Weather Report is drawn on a background which is white for the sea areas and pale green for the land areas, so that, like newspaper forecasts, all isobars and frontal systems are in black ink, the fronts appearing, as appropriate, in the tooth or dome notation with which we are familiar.

The original weather map, however, is a much prettier and more technical affair, in which certain conditions are entered in coloured pencil, and frontal divisions are denoted by colour and not by tooth-and-dome silhouettes. Let us see how these colours work out.

Station Plot	recorded in red and black ink for contrast and ease of reading.
Warm front	continuous red line.
Cold front	continuous blue line.
Occlusion	continuous purple line.
Warm occlusion	continuous thin red line behind a thin purple line.
Cold occlusion	continuous thin blue line behind a thin purple line.
Stationary occlusion	continuous purple line. Usually almost parallel to the isobars.
Stationary front	alternate red and blue lines formed in continuous succession.
Frontogenesis	corrugated lines of blue for warm front and red for cold front.
Frontolysis	short stroke through the front of the same colour as the front.
Precipitation	green shading.
Fog	yellow shading.

Well, it does seem rather a pity that we are not normally called upon to use such an interesting map, and also that we shall never be able to construct such a map for ourselves from the scant information available to us. Nevertheless, the matter of constructing our own map is not to be ended abruptly there, as we can, for the sake of a little listening time, provide ourselves with a reasonably up-to-date pressure map of the British Isles which can be surprisingly accurate.

Shipping Bulletins

The essence of good forecasting is to have information which is as fresh as possible so that little change has occurred since the observations were taken. We have seen that the Daily Weather Report is several hours behind being current information by the time it reaches us, and that newspaper forecasts suffer from the same deficiency. Radio and television bulletins are a little better but as the television weather comes rather late in the day, it is of little use to the day-to-day observer, and radio reports are necessarily of a scant nature.

The Shipping Bulletins broadcast by the BBC on 1500 metres (long wave) provide Sea Area and Coastal Reports, and a General Synopsis, all of which is read at normal reading speed and can be noted down easily by means of simple abbreviations and weather codes.

When that has been completed, you have a report of conditions around the British Isles which is less than 2 hours old and may be used to confirm any other backing information that you may have devised.

In order to facilitate the quick entry and subsequent production of a pressure map, the Royal Meteorological Society, publish the Metmap, which is designed solely for the Shipping Bulletin and is ideally laid out for its reception (Plate 27). The student should send for a pad of Metmaps without delay since their use will add greatly to his enjoyment and accuracy of future work.

Conveniently, the Metmap and Reports are adjacent to one another on the same page so that they could be easily mounted on a board under a clear plastic sheet, the entries and isobars being laid in with black marker. The sheet can be made ready for the next broadcast by a wipe over with a duster dipped in

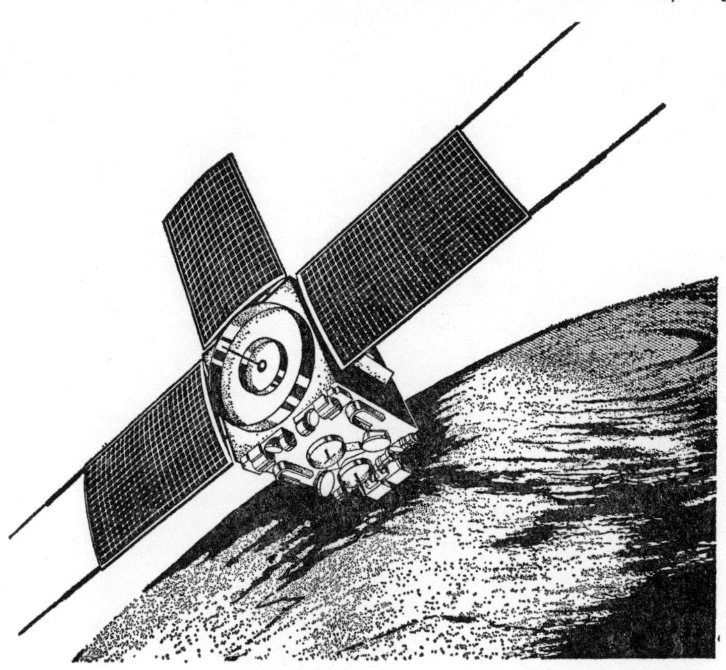

Fig. 25a *Artist's impression of a NOAA polar-orbiting satellite.*

meths. But if you prefer to keep your maps for comparisons later, then a biro or pencil will be suitable for use on the uncovered pad.

The most comprehensive instructions for receiving and dealing with these broadcasts are published by the Royal Meteorological Society in the form of a booklet by C. E. Wallington entitled *Your Own Weather Map,* which should be ordered at the same time as your Metmap.

Examination of the Metmap will show the exact location of those forecast areas whose names are familiar but whose position few of us know. Now is the time to catch up with Viking, Forties and Cromarty, Fastnet, Shannon, Rockall and Malin, and all the other vaguely romantic names of the seas around our coasts.

A panel in the lower right-hand corner of the Metmap shows examples of station plots, frontal lines and plotting symbols,

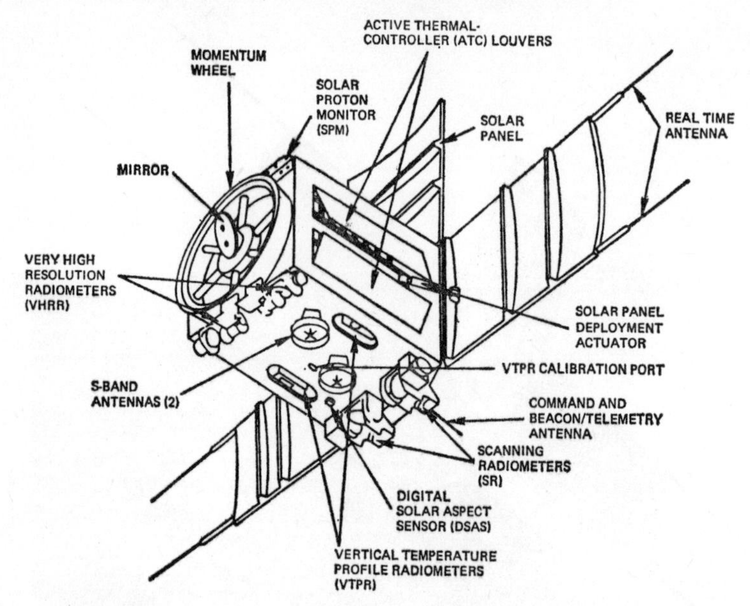

Fig. 25b *Schematic diagram showing the principal features of a NOAA satellite.*

and also provides a cut-out graduated scale which can be used for measuring wind strength from the distance between isobars.

In practice, the construction of a weather map from the information taken down in the report is not normally accomplished in one go, but necessitates the laying-in of isobars in a fairly sketchy manner, erasing and re-drawing them as a more logical pattern develops. So, to start with, put in the isobars with a black soft lead pencil which will rub out easily.

A good deal of concentration is needed to receive a Bulletin, since one has to listen to the reader, select the most relevant material and transpose it into abbreviations which can afterwards be interpreted into simplified station circles on the map.

The radio receiver should be switched on a couple of minutes before the Bulletin begins so that the station can be properly tuned and the ears become accustomed to the sound of broadcasting. When the Bulletin starts, it is important that it should be followed with dedication and without hesitation, otherwise subsequent sentences are easily missed. In other words, take it all down without trying to sort-out the sense of it as you go along. The time to interpret the details of the Bulletin is when they are being converted into plotting symbols on the map.

In construction, the broadcast is read in four basic parts:

i Gale warnings
ii General Synopsis
iii Sea Area Forecast
iv Coastal reports.

The Gale Warnings are indicated in the extreme left-hand column of the Report, next to the sea areas, and a gale is denoted by an X placed against the appropriate sea area name. So, if a gale is forecast for Sole, Lundy and Fastnet, an X is entered in each left-hand box. The announcement is quite simply to the effect that 'gale warnings are in operation in Sole, Lundy and Fastnet'.

The General Synopsis is located at the top of the page and also bears the time and date of the Bulletin.

This provides the frontal and pressure situation for the sea areas for the following 24 hours, and will mention the position and movement trends of depressions, anticyclones, troughs, ridges and fronts. There is no need to write down the full words, but use the following notation:

C = cold front
H = high pressure (anticyclone)
L = low pressure (depression)
O = occlusion
R = ridge of high pressure
T = trough of low pressure
W = warm front

The indicator L is always followed by the millibars for the depression, and the indicator H is always followed by the millibars for the anticyclone, so that one may write: L 980 or H 1016 in the column headed *System.*

The *Present Position* of the system is noted as being, perhaps '100 miles west of Shannon'; its *Movement* may be 'expected to move east', and its *Forecast position* may be expected 'to be centred in Fastnet with a central pressure of 984 millibars by midnight'. The time is entered under the last column which is headed *at*. Time starts from 00 GMT, being midnight at the beginning of the day, to 24, being midnight at the end of the same day.

The Sea Area Forecast follows, and is always given in the order shown on the report, starting with Viking and travelling clockwise around the British Isles, through Fisher, Biscay, Shannon, Bailey and Faroes, up to South-east Iceland.

When the details concern more than one station at a time, it is easy to bracket those stations together within the right-hand margin of the forecast column. The reader simply says: 'Viking, Forties, Cromarty. Wind in the west, force 5 at first, increasing to force 6 or 7 later.' This information is separated and briefed into the columns headed *at first* and *later*. The notation 'W5' is good enough to indicate that the wind is in the west at force 5, and a simple '6–7' immediately shows that it is expected to increase later to force 6 or 7. If accompanied by a change of direction, then include the compass bearing with the expected force, so that it may come out as 'S 6–7' for 'South, force 6 or 7'.

The *Weather* column is one which allows the use of our recently acquired abbreviations and symbols, such as a dot for rain, a comma for drizzle, an inverted triangle for showers, a star for snow, etc., and you may prefer to use these for practice On the other hand, there is a simple table of letters which you may find easier:

D = drizzle	M = mist	F = fog
R = rain	Z = haze	Fp = fog patches
S = snow	T = thunder	H = hail
P = showers	TP = thundery showers	Q = squall
s = slight	p = poor	i = intermittent
loc = locally	m = moderate	g = good
c = continuous	occ = occasional	h = heavy
pp = perhaps	LV = light and variable	cyc = cyclonic

The phrase 'moving to' can be indicated by an arrow.

The passing of time, such as 'rain at first followed by showers', or 'drizzle at first', or 'rain later', may be indicated by an oblique stroke between the elements of information, so that

S/P would mean 'snow at first followed by showers'
S/ would mean 'snow at first'
/S would mean 'snow later'
p/g would mean 'poor visibility becoming good'

In practice, you may decide to use a mixture of these letters and the more professional-looking symbols, but this is a matter of individual selection, and as your work is not likely to be held up for exhibition purposes, the object should be to use whatever method enables you to take down a report which is understandable to *you,* and one which embodies all the essential details of the broadcast.

The Coastal Reports give details of wind direction and force, weather, visibility, pressure and change of pressure, in that order, for the eleven coastal stations:

Wick, Bell Rock, Dowsing, Galloper, Royal Sovereign, Portland Bill, Scilly, Valentia, Ronaldsway, Prestwick, Tiree,

and since these observations are the most recently received weather conditions to be broadcast, they are probably the most important items in the Bulletin.

To assist in speed of writing, it is necessary to abbreviate all wind directions to the initial letters only; the force, to the appropriate one or two figures; the weather, to plotting symbols; the visibility, to 1, 2 or 3 figures; and the pressure to 2 figures. Changes are indicated by the attitude of a stroke.

For example, there is no need to indicate miles or yards after visibility distances, since it can be assumed that 5=5 miles and that 10=10 miles, while 3 or 4 figures, such as 200 or 2000 can be assumed to be in yards. If time is pressing and the reading speed is uncomfortably fast, such as is sometimes the case when extra material is being crammed into the time limits of the broadcast, it is possible to abbreviate the pressure readings to two figures, omitting the initial 9 for low readings and the 10 for higher readings, so that 1004 millibars will read 04, and 996 millibars will read 96. It will be apparent that we are not getting a reading of 904 millibars on the one hand, or 1096 millibars on the other, since our barometric readings are between 940 millibars when very low and 1050 millibars when very high, and that we average between 980 and 1030 millibars as common variations.

If it is more convenient, and there is time to do so, by all means enter the full-figure readings and omit the word millibars (mbs).

In the column for *change,* which indicates the tendency of the barometer, such as 'steady', 'falling' or 'rising', one can draw a short horizontal stroke, a stroke sloping downwards to the left or a stroke sloping upwards to the right respectively, the angle of inclination being steep for rapid changes and shallow for slow changes.

Further detail would only be tedious as the best way to learn the procedure is to listen to two or three broadcasts before actually atempting to take them down, so that the pattern and speed become familiar. It is obviously an advantage to practise and learn the plotting symbols so that they come to mind without the need to think about them when taking down a Bulletin, and before this can be attempted it is essential to provide yourself with a pad of Metmaps.

Assuming that we have a full Bulletin and that we *intend* to plot the information direct on to our map, we must start with the most up-to-date information namely the Coastal Reports.

Plotting the weather

It will be seen by reference to the map that the eleven coastal stations are indicated by station circles. These we fill in with all the information we have about them, starting with the wind direction and proceeding in the order given in the Bulletin (Plate 27).

Wind direction. This is indicated by a feathered arrow with its nose buried in the station circle and its tail feathers indicating the direction from which the wind is blowing. If there is no wind and the area is calm, then an outer circle is drawn around the station circle to denote a 'calm'.

Wind force. A whole barb, or full feather, indicates two forces in the Beaufort Scale, a half barb indicates one force. So a feather with 1 whole and 1 half barb gives us a reading of force 3, and 2½ barbs give us force 5. The feathers are drawn, for the sake of consistency, on the left-hand side of the direction arrow when viewed looking towards the station from the tip of the tail.

Weather. This is plotted on the left of the station circle at position 9 o'clock, and provides an excellent change for the use of our plotting symbols, a general selection of which appears on the map for guidance.

Visibility. Shown directly to the left of the weather symbol, or in that position if there is no weather symbol. This means one or two figure numbers for miles distance, or three or four figure numbers for yards distance.

Pressure. In position 1 o'clock on the station circle, enter the pressure either in the full number of millibars or in the abbreviation used when taking down the Bulletin, which ever is convenient.

Pressure change. Immediately below the pressure, draw the appropriate line for any change recorded.

A little study reveals the general behaviour of the winds, the average pressure throughout the area plotted, and, in fact, everything we need to know about conditions at the time of observation.

Examine now the weather systems described in the *General Synopsis* and locate the highs and lows, the troughs and ridges on the map. A front is drawn in using the usual tooth-and-dome signatures according to type and condition. By way of example, take our earlier detail from the General Synopsis and note that we had a low pressure system of 980 millibars 100 miles west of Shannon, expected to move east to Fastnet by midnight with a central pressure of 983 millibars. In order to indicate this condition, we can measure off 100 miles west of Shannon by using the scale at the foot of the map, and we can mark it with an X, alongside of which we can show the pressure. From that point we draw a line through Shannon into Fastnet, ending with an X and the second pressure reading of 983.

This tells us the present and forecast positions of the low-pressure area, and will assist in making our final decisions.

When all such information has been roughly charted, we start on the *Sea Area Forecast,* and follow the same routine, plotting the wind at first, and later the weather and the visibility.

In showing the wind conditions, it is a good plan to use a large arrow to indicate the 'wind at first', and a smaller arrow, buried in the same spot on the target, to indicate the 'wind later', the two being separated by a slight angle.

The weather features are then entered, making good use of the oblique stroke to denote present weather 'becoming' or 'later'.

We now have a very localised mini-forecast map, but we have nevertheless followed out the standard procedure of gathering weather information and translating it into station circles preparatory to drawing in the isobars.

The scale provided with the Metmap shows that lines drawn at $\frac{1}{4}$-inch separation can be used to represent isobars drawn at 2-millibar intervals, which is the scale to be used in this case. Having decided the scale on which to work, we must now attempt to sketch in the isobars according to the wind arrows and pressure readings we find on the map. At first it is necessary to make lines of no more than an inch or two in length, following the direction of the wind arrows and sharply changing direction if the arrows indicate such a change, because the angle at which the change takes place will most likely indicate a front.

It is good practice to establish one or more isobars, such as might occur in a depression or anticyclone, and whose barometric pressure is positively known, so that you can lay off from them the surrounding isobars at 2-millibar separation according to the scale provided.

Actually deciding on how the isobars will run is the most difficult part of the job, which demonstrates adequately the amount of skill needed by the professional meteorologist, but after a few early attempts and a good deal of erasing and re-drawing, a certain degree of proficiency can be achieved, and eventually recognisable patterns emerge all over the map which give a good reading of the conditions prevailing at the time of the Bulletin.

To actually make a forecast from the information now plotted, one has to try to envisage the behaviour of the charted conditions in, say, 2 or 3 hours' time. This is done by studying the direction of travel and the speeds of various systems and fronts, and constructing a chart which represents your opinion of what conditions will be like at the forecast time you have set. Your opinion is, of course, based upon your knowledge of the weather pattern and behaviour of weather instruments under the influence of various weather conditions, so it may be as well to study again the average patterns within depressions and anticyclones and their attendant surroundings.

Enough seems to have been said about this subject to enable a practical start to be made in producing a home-made weather map. We must, in conclusion, turn our attention to the home-based weather station.

The home-based weather station

A keen observer making meteorology his most absorbing hobby may be prepared to elaborate as much as possible with expensive precision equipment. He may want to use a barograph instead of a simple barometer, or a thermograph instead of a simple thermometer, or even a recording rain gauge and a sunshine recorder to support his enthusiasm. While every advantage is offered by expensive equipment, there are not many of us who can afford their luxury and so we are brought instead to the less costly and less elaborate instruments for our daily use.

It is necessary to have:

- i A barometer for measuring the air pressure which leads us to anticipating the kind of conditions approaching. This can be mounted indoors, but away from sunlight, radiators and excessive vibration.
- ii Wet and dry bulb thermometers for measuring the humidity.
- iii Maximum and minimum thermometers for registering the maxima and minima in the absence of the observer.
- iv Minimum thermometer for grass temperatures.
- v A wind vane to give us wind direction.
- vi A standard 5-inch rain gauge.

Apart from the barometer and grass thermometer, all these instruments should be mounted in a louvered screen sited on an exposed piece of ground where it is not influenced by heat radiation from buildings or fences. Not many garden plots conform to the ideal requirements of siting a thermometer screen, and so it might be found necessary to mount a box of thermometers in a spot which is protected from sunshine and wind, such as a north-facing wall.

In addition to the observations provided by instruments, there are, of course, the all important visually observed elements such as visibility, cloud amount, cloud type and type of precipitation.

It would not come amiss to have a small shed as a plotting and logging room, since it is then possible to hang all your daily, weekly and monthly charts on the walls for comparison. Furthermore, a copy of the Metmap can be fixed to a sheet of plywood and covered by a layer of clear plastic so that the weather situation can be marked on it with a felt-tipped marker and erased with meths when no longer required. Providing that the shed is insulated, a transistor radio could be kept there permanently tuned to the Shipping Bulletin on 1500 metres, ready to switch on when needed.

The question of exactly when weather observations should be taken must depend upon individual commitments, such as working pattern and holidays etc., but ideally the observation times should be 7 or 8 a.m., 1 p.m., 6 or 7 p.m. and midnight. Exactly how that routine is fitted in is a matter for individual concern, although it would seem that for those with normal working hours, the 7 a.m., and 7 p.m., observations are possible, while it may be necessary to advance the midnight one to say 10 or 11 p.m. The 1 p.m., reading could perhaps be taken by someone who is normally in the house at that time and who is familiar with the work to be carried out.

Obviously, for those who are at home all day no problems arise, except that it is essential for regular and accurate observations to be made. Any wide deviation will result in inconsistent reports.

The Weather Log should be made in a good quality, hard-cover, notebook, and should be entered on two pages, the left-hand page being divided into columns representing each reading to be taken, and the right-hand page devoted to other observations, such as the type of weather actually experienced, sky colour, rain, and everything that can contribute to the fullness of the observation.

Example: (left-hand page)

Observation Station......

	Barometer		Thermometers							
Date	7 a.m.	10 p.m.	7 a.m.	1 p.m.	6 p.m.	10 p.m.	Max	Min	Wind dir.	Speed
Jan 1	1016	1014	30	37	39	40	40	30	S	5 knots

Other columns could be allowed for humidity and dew point readings at each observation time.

The right-hand page can be set out in columns or simply left as free writing space into which all visual observations may go, but individual choice is best for personal comfort and ease of working.

To avoid the possibility of the log becoming dirty and damaged while attempting to note down readings in the rain or in half a gale, it is a good idea to use a rough notebook for this purpose, and to transfer the information to the log proper when it can be written up neatly and clearly.

There are many kinds of additional material which can be entered as time and circumstances allow; for example, the temperature of the garden pond, and the amount of evaporation from it measured against a plastic ruler. The temperature and humidity of a permanently shaded area can be taken, and a note made of the differences in temperature between the front and the back of the house.

There are also vertical variations of temperature between the ground floor and upper storey of the building, and horizontal changes from one side of the garden to the other. There are enormous differences in all readings in the special conditions existing within a greenhouse and beneath glass or plastic cloches.

Clearly, there are a good many ways in which to expand upon the basic work of making regular observations, and not the least of these is the study of local climatology and of how surrounding topography may affect the weather conditions experienced in certain areas of the county.

For further information on the subject of observing, it is worth obtaining from H.M.S.O. a copy of *The Observer's Handbook* and the *Handbook of Meteorological Instruments*, Part I.

There are a good many publications on weather and weather forecasting, and those recommended by the Royal Meteorological Society provide a basis for starting a weather library, an investment which is not necessarily expensive but is invaluable for the advancement of one's personal skills. In particular, no library is complete without a good cloud atlas, and one of the best with which to make a start is *A Colour Guide to Clouds* by Professor Richard Scorer and the late Dr Harry Wexler, obtainable from the Royal Meteorological Society along with their other publications.

Meteorological satellites

Since nothing has so extended the view of those who seek to monitor the physical atmosphere as the advent of earth-orbiting satellites, it will repay our curiosity to dwell a while in the sophisticated realm of meteorological satellites so that we may come to appreciate the extent of the subject.

Since the early explorer days of the 1960s, when scientists gained experience and operational data by way of the earth-orbiting TIROS series satellites, and since the Applications Technology Satellites (ATS) pioneered the automatic scanning of the atmosphere from space, man's ability to observe the Earth's environment from space has increased with surprising swiftness, and technological improvements now provide us with an operationally useable view of the behaviour of the weather elements in their continuing battle with each other.

In fact, the satellite has become one of the important ways in which man monitors the whole atmosphere, the oceans, the solid earth, the invisible forces of gravity and magnetism, and the intricate impact of solar storms upon our planetary home. And still the National Environmental Satellite Service of the U.S.A. is developing new ways of using this vantage point in space to improve our ability to understand, describe and predict conditions in the day-to-day physical environment.

The United States National Oceanic and Atmospheric Administration (NOAA) and the National Environmental Satellite Service (NESS) operate geostationary satellites in daily routine surveys of large portions of the world's weather as observed by electronic sensors from space orbit, and in this sphere of survey are developing new methods of using this space-located observation point to improve satellite meteorologists' ability to predict conditions in the atmosphere.

NESS is part of the World Meteorological Centre in Washington, and is one of three World centres established under the United Nations' World Meteorological Organisation, the others being in Melbourne and Moscow.

There are two types of satellites operated by the NESS, the polar-orbiting spacecraft called NOAA, and other satellites in geostationary orbit (or earth-synchronous orbit, as you wish) called GOES.

NOAA satellites, Figs 25a and b, occupy a nearly circular, nearly polar orbit about 900 miles high and provide a twice daily global coverage of surface temperatures, solar proton levels in space, temperature profiles from the surface upwards to about 19 miles, and a useful coverage of cloud patterns. The disadvantage of the system is that its scanners can 'see' a developing local storm only once in each twelve hours of the total day, and therefore, some critical developments in the condition of the atmosphere may be missed entirely, having appeared just after the scan has passed, and having done their damage and decayed long before the scanner again photographs that area of the globe. It is, then, apparent that in addition to the valuable information provided by the NOAA satellite, further and more continuous data is needed if we are to attempt a round-the-clock survey of the atmosphere, and this type of data is provided by near-continuous views of our planet's surface from a satellite orbiting in a position 22,300 miles in space.

The Geostationary Operational Environmental Satellite (GOES), Plate 26, which is also known as a Synchronous Meteorological satellite (SMS), is more sophisticated than its ancestors, and is designed to scan all of North and South America and adjacent ocean areas with good resolution in half-hourly transmissions.

'Why', we may ask, 'should these particular satellites occupy that particular position in space, and, on the other hand, what is an orbit?'

All the stars, planets, suns and moons are in orbit, occupying their own special place in the vast picture of the sky at night; and man can add his own factory-made 'planets' to those already in motion around and above us.

If a suitably designed object, such as a satellite or space rocket, can be propelled at sufficient speed in the right direction through the atmosphere and into space, it can be orbited at any altitude above it, remaining in space for ever or being capable of return having fulfilled its purpose. However, as the distance between satellite and planet increases, the speed required to maintain an orbit *decreases,* and at an altitude of about 22,300 miles, the orbital speed is down to about 6,800 m.p.h., and the period of the circular orbit becomes 24 hours.

If this 22,300 mile high orbit lies in the plane of the Earth's equator, the satellite and the Earth turn through the same arc distance in the same time, so that the satellite is always above the same point on the equator. This is what is meant by geostationary, or earth-synchronous orbit – when the angular rotation of a point on the earth and the satellite is the same – so that the satellite can 'mark time' above the same position on the equator, spin-stabilised with its spin axis parallel to the Earth's axis.

The GOES satellite's spin-scan radiometer can provide visible and infra-red observations of the Earth below every 30 minutes, day and night.

Changes in the geomagnetic field and the flow of energetic material from the Sun – electrons, protons and X-radiation – are sensed by its space environment monitors for onward transmission to centres on the Earth below. Two GOES spacecraft, in geostationary orbit over the equatorial Atlantic and Pacific can provide coverage of a large portion of the western hemisphere.

Scientists and electronic equipment at ground stations transpose satellite information into analyses of wind fields, cloud temperatures, interhemispheric mixing, data from ocean-buoys, weather maps, time-lapse films, changes in the space environment, and satellite photographs; a formidable, three-dimensional task and a marvel of scientific achievement which helps to provide some sort of perspective to our own individual contributions to weather forecasting.

Epilogue

A keen interest and enthusiasm are the essential factors in preparing oneself for weather observing, as well as an appreciation of the skill it needs to attempt a forecast in our changeable climate, a climate which, although well suited to the needs of the British people, is nevertheless a frequent target for derogatory remarks. A charming little satire is typical of the British attitude to their changeable weather:

> Dirty days hath September,
> April, June and November;
> From January up to May,
> The rain it raineth every day.
> All the rest have thirty-one,
> Without a blessed gleam of sun;
> And if any of them had two-and-thirty,
> They'd be just as wet and twice as dirty.

Fortunately, the overall picture is by no means as gloomy as this suggests – we do experience mild winters and good summers, and throughout the year we have a selection of cold days and warm days; days without sunshine and days with long periods of sunshine, but because it is often masked from view by the clouds of successive weather systems, we do not always observe the Sun direct. Nevertheless, it is always there, shining brightly in the eternal night of outer space – shining to keep our Earth alive and habitable – shining to provide an interesting domain for the weather forecaster and the writer, for if the Sun ever 'went out' there would be no need for anyone to write another book about the weather.

INDEX